Tuncer Cebeci

Turbulence Models and Their Application

HORIZONS
PUBLISHING

Long Beach, California
Heidelberg, Germany

Springer
Berlin   *London*
Heidelberg *Milan*
New York   *Paris*
Hong Kong   *Tokyo*

Tuncer Cebeci

Turbulence Models and Their Application

Efficient Numerical Methods
with Computer Programs

With 40 Figures and a CD-ROM

**HORIZONS
PUBLISHING**

Tuncer Cebeci

810 Rancho Drive
Long Beach, CA 90815, USA
TuncerC@aol.com

Library of Congress Cataloging-in-Publication Data
Cebeci, Tuncer
Turbulence models and their application: efficient numerical methods
with computer programs / Tuncer Cebeci.
p. cm. Includes bibliographical references and index.
ISBN 978-3-540-40288-6 (acid-free paper)
1. Turbulent boundary layer – Mathematical models. 2. Turbulence – Mathematical models.
I. Title. QA913.C425 2003 532′.0527′015118–dc21 2003014170

ISBN 978-3-540-40288-6 Springer-Verlag Berlin Heidelberg New York

Additional material to this book can be downloaded from http://extras.springer.com

Typeset in MS Word by the author. Edited and reformatted by Kurt Mattes, Heidelberg, Germany,
using LATEX.
Cover design: Erich Kirchner, Heidelberg, Germany

Printed on acid-free paper   5 4 3 2 1 0

Preface

Turbulence models are required to close forms of the time-independent Navier–Stokes equations, which involve unknown Reynolds stress correlations. The equations may be incompressible or compressible, two- or three-dimensional, appropriate to separated or boundary-layer flows and with correlations that involve from one shear stress to three shear stresses and three normal stresses: and the averaging may involve ensembles, time and density weighting. In all cases, the averaging implies more unknowns than equations and assumptions are necessary to close the set. An important consequence is that solutions of equations closed with turbulence models are no longer exact representations of physical problems implied by the boundary conditions, and the uncertainties associated with the assumptions have to be appraised, usually by comparison with experiments.

Many models have been suggested over the last forty years, with initial emphasis on comparatively simple free flows and wall flows in which the unknown Reynolds-stress tensor could be reduced to one dominant shear-stress component. The ability to measure the shear stress and related mean-velocity gradients implied that assumptions could be evaluated directly and often, but not always, with encouraging results. It has proved more difficult to make evaluations of similar quality in more complex flows, for example those involving three independent variables and for substantial regions of flow separation, for which the Reynolds stress tensor involves more terms, although the correspondingly more complex models involve assumptions and empirical coefficients derived by reference to simpler flows. Thus, the extent to which a turbulence model is able to represent a flow depends on the complexity involved; assessments can be difficult and involve subjective judgement. At the same time, the Reynolds-stress tensor is one group of terms in the equation representing conservation of mean momentum and its magnitude is, on occasion, small compared to the terms representing pressure gradients and sources.

The implementation of turbulence models involves the numerical solution of the conservation equations for mass and momentum with the model assumptions for turbulent diffusion, and some compatibility between equations and assumptions is required to ensure solutions with reasonable computer resources. Thus, it can be expected that models in algebraic form are likely to allow results to be obtained with less cost, effort and time than those with models in the form of added differential equations for turbulence quantities, and some tradeoff may be required between cost and convenience on the one hand and greater generality on the other.

In this book, after a brief review of the more popular turbulence models (Chap. 1), we present and discuss accurate and efficient numerical methods for solving the boundary-layer equations with turbulence models based on algebraic formulas (mixing length, eddy viscosity) or partial-differential transport equations. In Chap. 2 we discuss the numerical solution of the boundary-layer equations using the Cebeci–Smith model and the k-ε model with and without wall functions, including a zonal method, all for flows without separation. A computer program employing the Cebeci-Smith model and the k-ε model for obtaining the solution of two-dimensional incompressible turbulent flows is discussed in detail in Chap. 3 and is presented in the accompanying CD-ROM.

The numerical solution of the boundary-layer equations with flow separation is discussed in Chap. 4. An inverse boundary-layer method employing the Cebeci-Smith turbulence model is described in detail. A computer program for obtaining boundary-layer solutions on airfoils and wakes is discussed and presented in the accompanying CD-ROM. This code, with Veldman's interaction law, can also be used interactively with the Hess and Smith panel method described in Chap. 5. The panel method, also presented in the accompanying CD-ROM, includes the viscous effects. It is arranged in such a way that, with the computer program of Chap. 4, it can be used to obtain solutions of airfoil flows for a wide range of angles of attack, including stall, as discussed in Chap. 7. Chapter 6 discusses the application of the computer program for the Cebeci-Smith and k-ε models to other higher-order turbulence models, including flows with separation. Finally, in Chap. 7 test cases are presented for the four companion computer programs in the accompanying CD-ROM. In many respects, this book can be seen as complementary and supplementary to that of Cebeci and Smith *Analysis of Turbulent Boundary Layers* (Academic Press, 1974) and its replacement *Analysis of Turbulent Flows* by the present author which is about to be published (Elsevier, 2003).

It is a pleasure to acknowledge the help received from Dr. K. C. Chang, who read the whole manuscript and assisted with the development of the boundary-layer codes. Dr. J. P. Shao prepared the CD-ROM for the computer programs and Professor H. H. Chen provided help in the development of the panel code with viscous effects. Thanks are also due to Mr. Kurt Mattes who typed the manuscript and to Mr. Karl Koch who helped with the the production of the book.

Indian Wells, July, 2003   *Tuncer Cebeci*

Contents

1

Turbulence Models

1.0 Introduction

The use of Reynolds-averaged equations, made necessary by our inability to solve the time-dependent, three-dimensional Navier-Stokes equations with adequate resolution of time and spatial scales, see [1], implies that information has been lost and that further approximations are required to represent the fluctuating quantities known as Reynolds stresses and so reduce the number of unknowns to equal the number of equations. The most common approach to this problem is to define an eddy viscosity, ε_m, in the same form as the laminar viscosity. Thus, for a two-dimensional incompressible flow,

$$- \varrho \overline{u'v'} = \varrho \varepsilon_m \frac{\partial u}{\partial y} \tag{1.0.1}$$

Another approach is to use the mixing length, ℓ, concept and express the Reynolds shear stress by

$$- \varrho \overline{u'v'} = \varrho \ell^2 \left(\frac{\partial u}{\partial y} \right)^2 \tag{1.0.2}$$

The specification of ε_m or ℓ may be made in terms of algebraic equations or in terms of a combination of algebraic and differential equations and this has given rise to terminology involving the number of differential equations. Thus, the closures may be described in terms of zero, one and two differential equations. For a two-dimensional boundary-layer, the zero-equation approach usually treats a turbulent boundary layer as a composite layer with separate expressions for ε_m or ℓ in each region. The Cebeci-Smith (CS) model discussed in detail in [1, 2] and briefly in Section 1.1 is a typical example for this approach.

In the one-differential-equation approach the eddy viscosity is written, with c_μ denoting a constant, as

$$\varepsilon_m = c_\mu k^{1/2} \ell \tag{1.0.3}$$

with k obtained from a differential equation which represents the transport of turbulence energy and ℓ from an algebraic formula. The Spalart-Allmaras (SA) model discussed in detail in [1, 3, 4] and briefly in Section 1.5 is a good, useful model that uses this approach.

In the two differential equation approach, the eddy viscosity is written as

$$\varepsilon_m = \frac{c_\mu k^2}{\varepsilon} \tag{1.0.4}$$

with k and ε obtained from differential equations which represent the transport of turbulence energy and its rate of dissipation. While one-equation models have found little favor except for the SA model, and where transport of turbulence characteristics is important as in strong adverse gradients or in separated flows, two equations have found extensive use. Various forms of two-equation models have been proposed and details have been given, for example in [1, 4]. Three popular models that are based on this approach are the k-ε model discussed in Section 1.2 and the k-ω and SST models briefly discussed in Sections 1.3 and 1.4, respectively.

The Reynolds shear stress can also be modelled by using the Reynolds transport equation as described for example in [1, 4]. A popular model is due to Launder, Reece and Rodi [5]. As pointed out by Bradshaw "it is so obvious that stress-transport models are more realistic in principle than eddy viscosity models that the improvements they give are very disappointing and most engineers have decided that the increased numerical difficulties (complexity of programming, expense of calculation, occasional instability) do not warrant changing up from eddy-viscosity models at present. Even stress-transport models often give very poor predictions of complex flows – notoriously, the effects of streamline curvature are not naturally reproduced, and empirical fixes for this have not been very reliable" [6]. For this reason, the solution of the boundary layer equations using stress-transport models is not addressed in this book.

1.1 CS Model

For two-dimensional incompressible flows, the continuity and momentum equations are given by [2]

$$\frac{\partial u}{\partial x} + \frac{\partial v}{\partial y} = 0 \tag{1.1.1}$$

$$u \frac{\partial u}{\partial x} + v \frac{\partial u}{\partial y} = -\frac{1}{\varrho} \frac{dp}{dx} + \nu \frac{\partial^2 u}{\partial y^2} - \frac{\partial}{\partial y} (\overline{u'v'}) \tag{1.1.2}$$

With Bernoulli's equation, the above momentum equation can be written as

$$u\frac{\partial u}{\partial x} + v\frac{\partial u}{\partial y} = u_e\frac{du_e}{dx} + \nu\frac{\partial^2 u}{\partial y^2} - \frac{\partial}{\partial y}(\overline{u'v'}) \qquad (1.1.3)$$

The boundary conditions for Eqs. (1.1.1) and (1.1.3) are

$$y = 0, \qquad u = 0, \qquad v = 0, \qquad\qquad (1.1.4a)$$

$$y = \delta, \qquad u = u_e. \qquad\qquad (1.1.4b)$$

The Reynolds shear stress term in Eq. (1.1.3), that is, $-\varrho\overline{u'v'}$, requires a closure assumption. In the CS model this is achieved by using Eq. (1.0.1) with ε_m defined by separate expressions in the inner and outer regions of the boundary layer. Excluding the low-Reynolds-number and mass-transfer effects and assuming flow over a smooth surface (no roughness effects), the CS eddy viscosity model is given by [1].

Inner region: $0 \le y \le y_c$

$$(\varepsilon_m)_i = l^2\left|\frac{\partial u}{\partial y}\right|\gamma_{\text{tr}} \qquad\qquad (1.1.5)$$

Here the mixing length l is given by

$$l = \kappa y\left[1 - \exp\left(-\frac{y}{A}\right)\right] \qquad\qquad (1.1.6a)$$

where $\kappa = 0.40$ and A is a damping-length constant, which may be represented by

$$A = 26\frac{\nu}{N}u_\tau^{-1}, \quad N = (1 - 11.8p^+)^{1/2} \quad p^+ = \frac{\nu u_e}{u_\tau^3}\frac{du_e}{dx}, \qquad (1.1.6b)$$

Outer region: $y_c \le y \le \delta$

$$(\varepsilon_m)_0 = \alpha\int_0^\delta (u_e - u)dy\,\gamma_{\text{tr}}\gamma \qquad\qquad (1.1.7)$$

Here γ accounts for the intermittency of the outer region and is represented by

$$\gamma = \frac{1}{2}\left[1 - \text{erf}\frac{[y - Y]}{\sqrt{2}\,\sigma}\right] \qquad\qquad (1.1.8)$$

where Y and σ are general intermittency parameters, with Y denoting the value of y for which $\gamma = 0.5$, and σ the standard deviation. The dimensionless intermittency parameters Y/δ^* and σ/δ^* are expressed as functions of H as shown in Fig. 1.1a. The variation of the ratio of boundary-layer thickness δ to δ^* with H is shown in Fig. 1.1b. The parameter α is calculated from

$$\alpha = \frac{0.0168}{\left[1 - \beta\left(\frac{\partial u}{\partial x}\Big/\frac{\partial u}{\partial y}\right)_m\right]^{1.5}} \qquad\qquad (1.1.9)$$

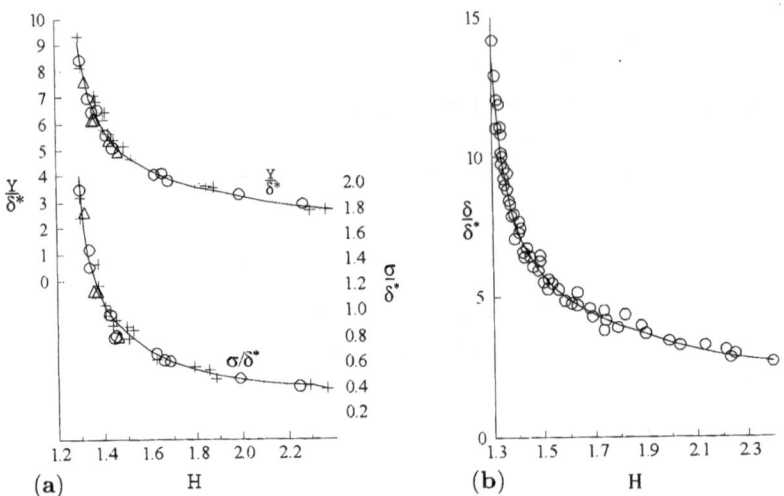

Fig. 1.1. Variation of Y/δ^*, σ/δ^* and δ/δ^* with H according to the data of Fiedler and Head [7].

Here subscript m denotes the location where turbulent shear is maximum. The parameter β is given by

$$
\beta =
\begin{cases}
\dfrac{6}{1 + 2R_t(2 - R_t)} & R_t \leq 1.0 \\[2ex]
\dfrac{1 + R_t}{R_t} & R_t \geq 1.0
\end{cases}
\tag{1.1.10}
$$

Here R_t denotes the ratio of wall shear to maximum Reynolds shear stress [1],

$$
R_t = \frac{\tau_\omega}{(-\varrho(\overline{u'v'}))_m}
\tag{1.1.11}
$$

In Eqs. (1.1.5) γ_{tr} is an intermittency factor which represents the streamwise region from the onset of transition to turbulent flow. It is defined by the following expression

$$
\gamma_{\mathrm{tr}} = 1 - \exp\left[-G(x - x_{\mathrm{tr}}) \int_{x_{\mathrm{tr}}}^{x} \frac{dx}{u_e}\right]
\tag{1.1.12}
$$

where x_{tr} is the location of the start of the transition and the factor G is given empirically by

$$
G = \frac{3}{C^2} \frac{u_e^3}{\nu^2} R_{x_{\mathrm{tr}}}^{-1.34}
\tag{1.1.13a}
$$

with $R_{x_{\mathrm{tr}}}$ denoting Reynolds number, $R_{x_{\mathrm{tr}}} = (u_e x/\nu)_{\mathrm{tr}}$, and C a constant with a recommended value of 60 for high Reynolds flows. For lower Reynolds number, C is given by [1]

$$
C^2 = 213(\log R_{x_{\mathrm{tr}}} - 4.7323)
\tag{1.1.13b}
$$

1.2 k-ε Model

The k-ε model due to Jones and Launder [8] is also based on the eddy viscosity concept, with ε_m given by Eq. (1.0.4). Here c_μ denotes a constant and k and ε are obtained from differential equations which represent the transport of turbulence kinetic energy, k, and its rate of dissipation, ε. They are given by [1, 4]

$$\frac{Dk}{Dt} = \frac{\partial}{\partial x_k}\left[\left(\nu + \frac{\varepsilon_m}{\sigma_k}\right)\frac{\partial k}{\partial x_k}\right] + \varepsilon_m\left(\frac{\partial \bar{u}_i}{\partial x_j} + \frac{\partial \bar{u}_j}{\partial x_i}\right)\frac{\partial \bar{u}_i}{\partial x_j} - \varepsilon \qquad (1.2.1)$$

$$\frac{D\varepsilon}{Dt} = \frac{\partial}{\partial x_k}\left[\left(\nu + \frac{\varepsilon_m}{\sigma_\varepsilon}\right)\frac{\partial \varepsilon}{\partial x_k}\right] + c_{\varepsilon 1}\frac{\varepsilon}{k}\varepsilon_m\left(\frac{\partial \bar{u}_i}{\partial x_j} + \frac{\partial \bar{u}_j}{\partial x_i}\right)\frac{\partial \bar{u}_i}{\partial x_j} - c_{\varepsilon 2}\frac{\varepsilon^2}{k} \qquad (1.2.2)$$

For boundary-layer flows at high Reynolds number, Eqs. (1.2.1) and (1.2.2) become

$$u\frac{\partial k}{\partial x} + v\frac{\partial k}{\partial y} = \frac{\partial}{\partial y}\left(\frac{\varepsilon_m}{\sigma_k}\frac{\partial k}{\partial y}\right) + \varepsilon_m\left(\frac{\partial u}{\partial y}\right)^2 - \varepsilon \qquad (1.2.3)$$

$$u\frac{\partial \varepsilon}{\partial x} + v\frac{\partial \varepsilon}{\partial y} = \frac{\partial}{\partial y}\left(\frac{\varepsilon_m}{\sigma_\varepsilon}\frac{\partial \varepsilon}{\partial y}\right) + c_{\varepsilon 1}\frac{\varepsilon}{k}\varepsilon_m\left(\frac{\partial u}{\partial y}\right)^2 - c_{\varepsilon 2}\frac{\varepsilon^2}{k} \qquad (1.2.4)$$

The parameters c_μ, $c_{\varepsilon 1}$, $c_{\varepsilon 2}$, σ_k and σ_ε are given by

$$c_\mu = 0.09, \quad c_{\varepsilon 1} = 1.44, \quad c_{\varepsilon 2} = 1.92, \quad \sigma_k = 1.0, \quad \sigma_\varepsilon = 1.3 \qquad (1.2.5)$$

These equations apply only to free shear flows. For wall boundary-layer flows, they require modifications to account for the presence of the wall. Without wall functions, it is necessary to replace the true boundary conditions at $y = 0$ by new "boundary conditions" defined at some distance y_0 outside the viscous sublayer to avoid integrating the equations through the region of large y gradients near the surface. Usually y_0 is taken to be

$$y_0 = \left(\frac{\nu}{u_\tau}\right)y_0^+,$$

y_0^+ being a constant taken as about 50 for smooth surfaces. For the velocity field, the boundary conditions at $y = y_0$ use the law of the wall and require that

$$u_0 = u_\tau\left(\frac{1}{\kappa}\ln\frac{y_0 u_\tau}{\nu} + c\right), \qquad (1.2.6a)$$

$$v_0 = -\frac{u_0 y_0}{u_\tau}\frac{du_\tau}{dx} \qquad (1.2.6b)$$

Here c is a constant about 5 to 5.2. Equation (1.2.6b) results from integrating the continuity equation with u given by the law of the wall

$$u^+ \equiv \frac{u}{u_\tau} = \phi_1(y^+) \qquad (1.2.7)$$

We also use relations for the changes in shear stress between $y = 0$ and $y = y_0$ in order to calculate u_τ from

$$u_\tau^2 = \frac{\tau_0}{\varrho} - \alpha y_0 \qquad (1.2.8a)$$

where τ_0 is calculated from

$$\tau_0 = \left[(\nu + \varepsilon_m) \frac{\partial u}{\partial y} \right]_{y_0}$$

with α semi-empirically given by

$$\alpha = 0.3 \frac{du_0^2}{dx} - u_e \frac{du_e}{dx} \qquad (1.2.8b)$$

The friction velocity u_τ $(\equiv \sqrt{\tau_w/\varrho})$ can also be calculated from

$$\tau = \tau_w + \frac{dp}{dx} y + \nu \frac{du_\tau}{dx} \int_0^{y^+} \left(\frac{u}{u_\tau} \right)^2 dy^+ \qquad (1.2.8c)$$

with u/u_τ determined from Thompson's velocity profile given by

$$u^+ = \begin{cases} y^+, & y^+ \leq 4 \\ c_1 + c_2 \ln y^+ + c_3 (\ln y^+)^2 + c_4 (\ln y^+)^3, & 4 < y^+ < 50 \end{cases} \qquad (1.2.9)$$

where $c_1 = 1.0828$, $c_2 = -0.414$, $c_3 = 2.2661$, $c_4 = -0.324$.

For $y^+ > 50$, we can use the logarithmic velocity profile, [1, 2].

$$u^+ = \frac{1}{\kappa} \ln y^+ + c \qquad (1.2.10)$$

where $\kappa = 0.41$ and $c = 5.0$.

There are several ways to specify the "wall" boundary conditions for k and ε. A common one for k makes use of the relation between shear stress τ and k. It is given by Bradshaw [1].

$$y = y_0, \qquad k_0 = \frac{\tau_0}{a_1} \qquad (1.2.11)$$

where $a_1 = 0.30$. With τ_0 defined by Eq. (1.0.1) and ε_m by Eq. (1.0.4), Eq. (1.2.11) becomes

$$a_1 = c_\mu \frac{k_0}{\varepsilon_0} \left(\frac{\partial u}{\partial y} \right)_0 \qquad (1.2.12)$$

The boundary condition for ε can be obtained by equating the eddy viscosity given by the CS model, $(\varepsilon_m)_{CS}$, to the eddy viscosity definition used in the k-ε model, Eq. (1.0.4) which, with low Reynolds number correction, can be written as

$$(\varepsilon_m)_{CS} = c_\mu f_\mu \frac{k^2}{\varepsilon} \qquad (1.2.13)$$

Here f_μ is a specified function discussed later in this section. Thus,

$$y = y_0, \qquad l_0^2 \left(\frac{\partial u}{\partial y} \right)_0 = c_\mu f_\mu \frac{k_0^2}{\varepsilon_0} \qquad (1.2.14)$$

where l is given by Eq. (1.1.6a).

The edge boundary conditions for the k-ε model equations, aside from the edge boundary condition for the momentum equation,

$$y \to \delta, \qquad u \to u_e(x) \qquad (1.2.15)$$

are

$$y \to \delta, \qquad k \to k_e, \qquad \varepsilon \to \varepsilon_e \qquad (1.2.16)$$

To avoid numerical problems, k_e and ε_e should not be zero. In addition, k_e and ε_e cannot be prescribed arbitrarily because their development is governed by the transport equations (1.2.3) and (1.2.4) written at the boundary-layer edge,

$$u_e \frac{dk_e}{dx} = -\varepsilon_e \qquad (1.2.17a)$$

$$u_e \frac{d\varepsilon_e}{dx} = -c_{\varepsilon 2} \frac{\varepsilon_e^2}{k_e} \qquad (1.2.17b)$$

The above equations can be integrated with respect to x with initial conditions corresponding to k_{e_0} and ε_{e_0} at x_0. The solution provides the evolutions of $k(x)$ and $\varepsilon(x)$ as boundary conditions for the k- and ε-equations.

Low-Reynolds-Number Effects

To account for the presence of the wall, it is necessary to include low-Reynolds-number effects into the k-ε model. Without such modifications, this model fails to predict the sharp peak in turbulence kinetic energy close to the surface for pipe and channel flow as well as fails to predict a realistic value of the additive constant c in the law of the wall.

There are several approaches that can be used to model Eqs. (1.2.3) and (1.2.4) near the wall region. For an excellent review of these models, see Wilcox [4] and Patel et al. [9].

Before we discuss the models considered here, it is useful to write the k-ε model equations in the following general form,

$$u \frac{\partial k}{\partial x} + v \frac{\partial k}{\partial y} = \frac{\partial}{\partial y} \left[\left(\nu + \frac{\varepsilon_m}{\sigma_k} \right) \frac{\partial k}{\partial y} \right] + \varepsilon_m \left(\frac{\partial u}{\partial y} \right)^2 - (\tilde{\varepsilon} + D) + F \qquad (1.2.18)$$

$$u \frac{\partial \tilde{\varepsilon}}{\partial x} + v \frac{\partial \tilde{\varepsilon}}{\partial y} = \frac{\partial}{\partial y} \left[\left(\nu + \frac{\varepsilon_m}{\sigma_\varepsilon} \right) \frac{\partial \tilde{\varepsilon}}{\partial y} \right] + c_{\varepsilon 1} f_1 \frac{\tilde{\varepsilon}}{k} \varepsilon_m \left(\frac{\partial u}{\partial y} \right)^2 - c_{\varepsilon 2} f_2 \frac{\tilde{\varepsilon}^2}{k} + E \qquad (1.2.19)$$

where D, E and F as well as σ_k, σ_ε, $c_{\varepsilon 1}$, $c_{\varepsilon 2}$, f_1, f_2 are model dependent and

$$\bar{\varepsilon} = \varepsilon - D \tag{1.2.20}$$

Of the several models to account for low-Reynolds-number effects, we consider the models of Chien [10] and Hwang and Lin [11] in the computer program given in Chapter 4. Other models discussed in [1, 4] can also be incorporated into this computer program.

The wall boundary conditions for Eqs. (1.2.18) and (1.2.19) are

$$y = 0, \qquad k = 0, \qquad \bar{\varepsilon} = 0 \tag{1.2.21}$$

According to Chien's model, the parameters D, E, F, f_μ, f_1, f_2, σ_k, σ_ε, c_{ε_1} and c_{ε_2} are given by

$$D = 2\nu \frac{k}{y^2}, \quad E = -2\nu \left(\frac{\bar{\varepsilon}}{y^2}\right) \exp\left(-\frac{1}{2} y^+\right), \quad F = 0$$

$$f_\mu = 1 - \exp(-0.0115 y^+), \quad f_1 = 1.0, \quad f_2 = 1 - 0.22 \exp\left(-\frac{R_T^2}{36}\right) \tag{1.2.22a}$$

$$\sigma_k = 1.0, \quad \sigma_\varepsilon = 1.3, \quad c_{\varepsilon_1} = 1.35, \quad c_{\varepsilon_2} = 1.8$$

where

$$R_T = \frac{k^2}{\bar{\varepsilon}\nu} \tag{1.2.22b}$$

According to Hwang and Lin's model, the parameters in Eqs. (1.2.18) and (1.2.19) are given by

$$D = 2\nu \left(\frac{\partial \sqrt{k}}{\partial y}\right)^2, \qquad E = -\frac{\partial}{\partial y}\left(\nu \frac{\bar{\varepsilon}}{k} \frac{\partial k}{\partial y}\right),$$

$$F = -\frac{1}{2} \frac{\partial}{\partial y}\left(\nu \frac{k}{\bar{\varepsilon} + D} \frac{\partial D}{\partial y}\right), \qquad f_\mu = 1 - \exp(-0.01 y_\lambda - 0.008 y_\lambda^3),$$

$$f_1 = 1.0, \qquad f_2 = 1.0, \qquad \sigma_k = 1.4 - 1.1 \exp[-(y_\lambda/10)], \tag{1.2.23}$$

$$\sigma_\varepsilon = 1.3 - \exp[-(y_\lambda/10)], \qquad c_{\varepsilon_1} = 1.44, \qquad c_{\varepsilon_2} = 1.92,$$

$$y_\lambda = \frac{y}{\sqrt{\nu k/\bar{\varepsilon}}}$$

Another approach to include the low-Reynolds-number effects in the k-ε model is to employ a simpler model near the wall (a mixing-length model [12] or a one equation model [13] which is valid only near the wall region) and the full two-equation model in the outer region of the boundary layer; the two solutions are matched at a certain point in the boundary layer. This approach is sometimes referred to as the two-layer method or zonal approach. We shall use the latter name as we discuss the solution of the k-ε models with the CS model near the wall region.

1.3 *k-ω* Model

Like the *k-ε* model discussed in the previous section, the *k-ω* model is also popular and widely used. Over the years, this model has gone over many changes and improvements as described in [4]. The most recent model is due to Wilcox [4] and is given by the following defining equations.

With ε_m defined by

$$\varepsilon_m = \frac{k}{\omega} \tag{1.3.1}$$

the turbulence kinetic energy and specific dissipation rate equations are

$$\frac{Dk}{Dt} = \frac{\partial}{\partial x_k}\left[\left(\nu + \frac{\varepsilon_m}{\sigma_k}\right)\frac{\partial k}{\partial x_k}\right] + R_{ik}\frac{\partial u_i}{\partial x_k} - \beta^* k\omega \tag{1.3.2}$$

$$\frac{D\omega}{Dt} = \frac{\partial}{\partial x_k}\left[\left(\nu + \frac{\varepsilon}{\sigma_\omega}\right)\frac{\partial \omega}{\partial x_k}\right] + \alpha\frac{\omega}{k}R_{ik}\frac{\partial u_i}{\partial x_k} - \beta\omega^2 \tag{1.3.3}$$

where R_{ik} is given by

$$R_{ik} = \varepsilon_m\left(\frac{\partial u_i}{\partial x_k} + \frac{\partial u_k}{\partial x_i}\right) \tag{1.3.4}$$

and

$$\alpha = \frac{13}{25}, \quad \beta = \beta_0 f_\beta, \quad \beta^* = \beta_0^* f_\beta, \quad \sigma_k = 2, \quad \sigma_\omega = 2 \tag{1.3.5a}$$

$$\beta_0 = \frac{9}{125}, \quad f_\beta = \frac{1 + 70\chi_\omega}{1 + 80\chi_\omega}, \quad \chi_\omega = \left|\frac{\Omega_{ij}\Omega_{jk}S_{ki}}{(\beta_0^*\omega)^3}\right| \tag{1.3.5b}$$

$$\beta_0^* = \frac{9}{100}, \quad f_\beta = \begin{cases} 1, & \chi_k \le 0 \\ \dfrac{1 + 680\chi_k^2}{1 + 400\chi_k^2}, & \chi_k > 0 \end{cases}, \quad \chi_k = \frac{1}{\omega^3}\frac{\partial k}{\partial x_j}\frac{\partial \omega}{\partial x_j} \tag{1.3.5c}$$

The tensors Ω_{ij} and S_{ki} appearing in Eq. (1.3.5b) are the mean rotation and mean-strain-rate tensors, respectively, defined by

$$\Omega_{ij} = \frac{1}{2}\left(\frac{\partial u_i}{\partial x_j} - \frac{\partial u_j}{\partial x_i}\right), \quad S_{ki} = \frac{1}{2}\left(\frac{\partial u_k}{\partial x_i} + \frac{\partial u_i}{\partial x_k}\right) \tag{1.3.6}$$

The parameter χ_ω is zero for two-dimensional flows. The dependence of β on χ_ω has a significant effect for round and radial jets [4]. This model takes the length scale in the eddy viscosity as

$$l = \frac{\sqrt{k}}{\omega} \tag{1.3.7a}$$

and calculates dissipation ε from

$$\varepsilon = \beta^*\omega k \tag{1.3.7b}$$

Wilcox's model equations have the advantage over the k-ε model that they can be integrated through the viscous sublayer, without using damping functions. At the wall the turbulent kinetic energy k is equal to zero. The specific dissipation rate can be specified in two different ways. One possibility is to force ω to fulfill the solution of Eq. (1.3.3) as the wall is approached [4]:

$$\omega \rightarrow \frac{6\nu}{\beta y^2} \text{ as } y \rightarrow 0 \tag{1.3.8}$$

The other [14] is to specify a value for ω at the wall which is larger than

$$\omega_w > 100\Omega_w$$

where Ω_w is the mean vorticity at the wall.

Menter [14] applied the condition of Eq. (1.3.8) for the first five grid points away from the wall (these points were always below $y^+ = 5$). He repeated some of his computations with $\omega_w = 1000\Omega_w$ and obtained essentially the same results. He points out that the second condition is much easier to implement and does not involve the normal distance from the wall.

The choice of freestream values for boundary-layer flows is

$$\omega_\infty > \lambda\frac{u_\infty}{L}, \qquad (\varepsilon_m)_\infty < 10^{-2}(\varepsilon_m)_{\text{max}}, \qquad k_\infty = (\varepsilon_m)_\infty\omega_\infty \tag{1.3.9}$$

where L is the approximate length of the computational domain and u_∞ is the characteristic velocity. The factor of proportionality $\lambda = 10$ has been recommended [14].

For boundary-layer flows, Eq. (1.3.2) reduces to Eq. (1.2.18) with

$$D = 0, \qquad \tilde{\varepsilon} = 0.09\omega k \tag{1.3.10}$$

The specific dissipation rate equation, Eq. (1.3.3) becomes

$$u\frac{\partial \omega}{\partial x} + v\frac{\partial \omega}{\partial y} = \frac{\partial}{\partial y}\left[\left(\nu + \frac{\varepsilon_m}{\sigma_\omega}\right)\frac{\partial \omega}{\partial y}\right] + \alpha\left(\frac{\partial u}{\partial y}\right)^2 - \beta_0\omega^2 \tag{1.3.11}$$

1.4 SST Model

The SST model of Menter [14] combines several desirable elements of existing two-equation models. The two major features of this model are a zonal weighting of model coefficients and a limitation on the growth of the eddy viscosity in rapidly strained flows. The zonal modeling uses the 1993 version of Wilcox's k-ω model near solid walls and Launder and Sharma's k-ε model [1, 15] near boundary layer edges and in free shear layers. This switching is achieved with a blending function of the model coefficients. The shear stress transport (SST) modeling also modifies the eddy viscosity by forcing the turbulent shear stress to be bounded by a constant times the local turbulent kinetic energy. This modification, which is similar to the basic idea behind the Johnson-King model [16], improves the prediction of flows with strong adverse pressure gradients and separation.

In order to blend the k-ω model and the k-ε model, the latter is transformed into a k-ω formulation. The differences between this formulation and the original k-ω model of 1993 are that an additional cross-diffusion term appears in the ω-equation and that the modeling constants are different. Some of the parameters appearing in the k-ω model are multiplied by a function F_1 and some of the parameters in the transformed k-ε model by a function $(1 - F_1)$ and the corresponding equations of each model are added together. The function F_1 is designed to be a value of one in the near wall region (activating the original model) and zero far from the wall. The blending takes place in the wake region of the boundary layer.

The SST model also modifies the turbulent eddy viscosity function to improve the prediction of separated flows. Two-equation models generally underpredict the retardation and separation of the boundary layer due to adverse pressure gradients. This is a serious deficiency, leading to an underestimation of the effects of viscous-inviscid interaction which generally results in too optimistic performance estimates for aerodynamic bodies. The reason for this deficiency is that two-equation models do not account for the important effects of transport of the turbulent stresses. The Johnson-King model [16] has demonstrated that significantly improved results can be obtained with algebraic models by modeling the transport of the shear stress as being proportional to that of the turbulent kinetic energy. A similar effect is achieved in the SST model by a modification in the formulation of the eddy viscosity using a blending function F_2 in boundary layer flows [4, 14].

In the SST model, the eddy viscosity expression, Eq. (1.3.1), is modified,

$$\varepsilon_m = \frac{a_1 k}{\max(a_1\omega, \Omega F_2)} \tag{1.4.1}$$

where $a_1 = 0.31$. In turbulent boundary layers, the maximum value of the eddy viscosity is limited by forcing the turbulent shear stress to be bounded by the turbulent kinetic energy times a_1, see Eq. (1.2.11). This effect is achieved with an auxiliary function F_2 and the absolute value of the vorticity Ω. The function F_2 is defined as a function of wall distance y as

$$F_2 = \tanh(\text{arg}_2^2) \tag{1.4.2a}$$

where

$$\text{arg}_2 = \max\left(2\frac{\sqrt{k}}{0.09\omega y}; \frac{500\nu}{y^2\omega}\right) \tag{1.4.2b}$$

The two transport equations of the model for compressible flows are defined below with a blending function F_1 for the model coefficients of the original ω and ε model equations.

$$\frac{D\varrho k}{Dt} = \frac{\partial}{\partial x_k}\left[(\mu + \sigma_k \varrho \varepsilon_m)\frac{\partial k}{\partial x_k}\right] + R_{ik}\frac{\partial \bar{u}_i}{\partial x_k} - \beta^* \varrho \omega k \tag{1.4.3}$$

$$\frac{D\varrho\omega}{Dt} = \frac{\partial}{\partial x_k}\left[(\mu + \sigma_\omega \varrho \varepsilon_m)\frac{\partial \omega}{\partial x_k}\right] + \frac{\gamma}{\varepsilon_m} R_{ik}\frac{\partial \bar{u}_i}{\partial x_k} - \beta \varrho \omega^2$$
$$+ 2(1 - F_1)\varrho\sigma_{\omega 2}\frac{1}{\omega}\frac{\partial k}{\partial x_k}\frac{\partial \omega}{\partial x_k} \tag{1.4.4}$$

where

$$R_{ik} = \varrho \varepsilon_m \left(\frac{\partial \bar{u}_i}{\partial x_k} + \frac{\partial \bar{u}_k}{\partial x_i} - \frac{2}{3}\frac{\partial u_j}{\partial x_j}\delta_{ik}\right) - \frac{2}{3}\varrho k \delta_{ik} \tag{1.4.5}$$

The last term in Eq. (1.4.4) represents the cross-diffusion (CD) term that appears in the transformed ω-equation from the original ε-equation. The function F_1 is designed to blend the model coefficients of the original k-ω model in boundary layer zones with the transformed k-ε model in free shear layer and freestream zones. This function takes the value of one on no-slip surfaces and near one over a larger portion of the boundary layer, and goes to zero at the boundary layer edge. This auxiliary blending function F_1 is defined as

$$F_1 = \tanh(\text{arg}_1^4) \tag{1.4.6}$$

$$\text{arg}_1 = \min\left[\max\left(\frac{\sqrt{k}}{0.09\omega y}; \frac{500\nu}{y^2\omega}\right); \frac{4\varrho\sigma_{\omega 2}k}{\text{CD}_{k\omega}y^2}\right] \tag{1.4.7}$$

where $\text{CD}_{k\omega}$ is the positive portion of the cross-diffusion term of Eq. (1.4.4):

$$\text{CD}_{k\omega} = \max\left(2\varrho\sigma_{\omega 2}\frac{1}{\omega}\frac{\partial k}{\partial x_k}\frac{\partial \omega}{\partial x_k}, 10^{-20}\right) \tag{1.4.8}$$

The constants of the SST model are

$$\beta^* = 0.09, \qquad \kappa = 0.41 \tag{1.4.9}$$

The model coefficients β, γ, σ_k and σ_ω, denoted with the symbol ϕ, are defined by blending the coefficients of the original k-ω model, denoted as ϕ_1, with those of the transformed k-ε model, denoted as ϕ_2.

$$\phi = F_1\phi_1 + (1 - F_1)\phi_2 \tag{1.4.10}$$

where

$$\phi = \{\sigma_k, \sigma_\omega, \beta, \gamma\}$$

with the coefficients of the original models defined as
    inner model coefficients

$$\sigma_{k_1} = 0.85, \qquad \sigma_{\omega_1} = 0.5, \qquad \beta_1 = 0.075$$

$$\gamma_1 = \frac{\beta_1}{\beta^*} - \sigma_{\omega_1}\frac{\kappa^2}{\sqrt{\beta^*}} = 0.553 \tag{1.4.11}$$

outer model coefficients

$$\sigma_{k_2} = 1.0, \qquad \sigma_{\omega_2} = 0.856, \qquad \beta_2 = 0.0828$$

$$\gamma_2 = \frac{\beta_2}{\beta^*} - \frac{\sigma_{\omega_2}\kappa^2}{\sqrt{\beta^*}} = 0.440 \tag{1.4.12}$$

The boundary conditions of the SST model equations are the same as those described in the previous section for the k-ω model.

For incompressible boundary-layer flows, Eq. (1.4.3) reduces to the kinetic energy equation given by Eqs. (1.2.18) and (1.3.2). Equation (1.4.4) is same as Eq. (1.3.11) except that its right-hand side contains the cross diffusion term,

$$+ 2(1 - F_1)\sigma_{\omega_2}\frac{1}{\omega}\frac{\partial k}{\partial y}\frac{\partial \omega}{\partial y} \tag{1.4.13}$$

Two-equation models are not entirely reliable in complex flows. The "v2f" model of Durbin [17] gives generally better results, at the expense of using one more transport equation, nominally for the $-\varrho\overline{v'^2}$ Reynolds stress, and an elliptic relaxation equation to represent the effect of pressure fluctuations.

1.5 SA Model

Unlike the Cebeci-Smith model which uses algebraic expressions for eddy viscosity, this model uses a semi-empirical transport equation for eddy viscosity. Its defining equations are as follows.

$$\varepsilon_m = \tilde{\nu}_t f_{v1} \tag{1.5.1}$$

$$\frac{D\tilde{\nu}_t}{Dt} = c_{b_1}\left[1 - f_{t_2}\right]\tilde{S}\tilde{\nu}_t - \left(c_{w_1}f_w - \frac{c_{b_1}}{\kappa^2}f_{t_2}\right)\left(\frac{\tilde{\nu}_t}{d}\right)^2 + \frac{1}{\sigma}\frac{\partial}{\partial x_k}\left[(\nu + \tilde{\nu}_t)\frac{\partial \tilde{\nu}_t}{\partial x_k}\right]$$

$$+ \frac{c_{b_2}}{\sigma}\frac{\partial \tilde{\nu}_t}{\partial x_k}\frac{\partial \tilde{\nu}_t}{\partial x_k} \tag{1.5.2}$$

Here

$$c_{b_1} = 0.1355, \quad c_{b_2} = 0.622, \quad c_{\nu_1} = 7.1, \quad \sigma = \frac{2}{3} \tag{1.5.3a}$$

$$c_{w_1} = \frac{c_{b_1}}{\kappa^2} + \frac{(1 + c_{b_2})}{\sigma}, \qquad c_{w_2} = 0.3, \qquad c_{w_3} = 2, \qquad \kappa = 0.41 \tag{1.5.3b}$$

$$f_{v1} = \frac{\chi^3}{\chi^3 + c_{\nu_1}^3}, \qquad f_{v2} = 1 - \frac{\chi}{1 + \chi f_{v_1}}, \qquad f_w = g\left[\frac{1 + c_{w3}^6}{g^6 + c_{w3}^6}\right]^{1/6} \tag{1.5.3c}$$

$$\chi = \frac{\tilde{\nu}_t}{\nu}, \qquad g = r + c_{w_2}(r^6 - r), \qquad r = \frac{\tilde{\nu}_t}{\tilde{S}\kappa^2 d^2} \tag{1.5.3d}$$

$$\tilde{S} = S + \frac{\tilde{\nu}_t}{\kappa^2 d^2} f_{\nu_2}, \qquad S = \sqrt{2\Omega_{ij}\Omega_{ij}} \tag{1.5.3e}$$

$$f_{t2} = c_{t3} e^{-c_{t4}\chi^2}, \qquad c_{t3} = 1.1, \qquad c_{t4} = 2 \tag{1.5.3f}$$

where d is the distance to the closest wall and S is the magnitude of the vorticity, $\Omega_{ij} = \frac{1}{2}\left(\frac{\partial u_i}{\partial x_j} - \frac{\partial u_j}{\partial x_i}\right)$.

The wall boundary condition is $\tilde{\nu}_t = 0$. In the freestream and as initial condition 0 is best, and values below $\frac{\nu}{10}$ are acceptable [3].

For boundary-layer flows, Eq. (1.5.2) can be written as

$$u\frac{\partial \tilde{\nu}_t}{\partial x} + v\frac{\partial \tilde{\nu}_t}{\partial y} = c_{b_1}\left(1 - f_{t2}\right)\tilde{S}\tilde{\nu}_t + \frac{1}{\sigma}\left\{\frac{\partial}{\partial y}\left[(\nu + \tilde{\nu}_t)\frac{\partial \tilde{\nu}_t}{\partial y}\right] + c_{b_2}\left(\frac{\partial \tilde{\nu}_t}{\partial y}\right)^2\right\}$$

$$- \left(c_{w_1}f_w - \frac{c_{b_1}}{\kappa^2}f_{t2}\right)\left(\frac{\tilde{\nu}_t}{d}\right)^2 \tag{1.5.4}$$

where

$$\tilde{S} = \left|\frac{\partial u}{\partial y}\right| + \frac{\tilde{\nu}_t}{\kappa^2 d^2} f_{\nu_2} \tag{1.5.5}$$

References

[1] Cebeci, T.: Analysis of Turbulent Flows, Elsevier, 2003.

[2] Cebeci, T. and Cousteix, J.: Modeling and Computation of Boundary-Layer Flows, Horizons Publishing, Long Beach, CA and Springer, Heidelberg, Germany 1998.

[3] Spalart, P. R. and Allmaras, S. R.: A One-Equation Turbulence Model for Aerodynamics Flows. AIAA Paper 92-0439, 1992.

[4] Wilcox, D. C.: *Turbulence Modeling for CFD*. DCW Industries, Inc., 5354 Palm Drive, La Cañada, Calif., 1998.

[5] Launder, B. E., Reece, G. J. and Rodi, W.: Progress in the Development of a Reynolds Stress Turbulence Closure. *J. Fluid Mech.* **20**, 3, 1975.

[6] Bradshaw, P.: The Best Turbulence Models for Engineers. Modeling of Complex Flows (M. D. Salas, J. N. Hefner and L. Sakell, eds.) Kluwer, Dordrecht, 1999.

[7] Fiedler, H. and Head, M. R., "Intermittency Measurements in the Turbulent Boundary Layer," *J. Fluid Mech.* **25**, 719–735, 1986.

[8] Jones, W. P. and Launders, B. E.: The Predicition of Laminarization with a Two-Equation Model of Turbulence. *Int. J. Heat and Mass Transfer* **15**, 301–314, 1972.

[9] Patel, V. C., Rodi, W., and Scheuerer, G.: Turbulence Models for Near-Wall and Low Reynolds Number Flows: A Review. *AIAA J.* **23**, 1308–1319, 1985.

[10] Chien, K. Y.: Predictions of Channel and Boundary-Layers Flows with a Low-Reynolds-Number Turbulence Model. *AIAA J.* **20**, 33–38, 1982.

[11] Hwang, C. B. and Lin, C. A.: Improved-Low-Reynolds Number k-$\tilde{\varepsilon}$ Model Based on Direct Numerical Simulation Data. *AIAA J.* **36**, 38–43, 1998.

[12] Arnal, D., Cousteix, J., and Michel, R.: Couche limite se développant avec gradient de pression positif dans un écoulement turbulent. *La Rech. Aérosp.* 1976-1, 1976.

[13] Norris, L. H. and Reynolds, W. C.: Turbulent Channel Flow with a Moving Wavy Boundary. Report FM-20, Department of Mechanical Engineering, Stanford University, Stanford, California, 1975.

[14] Menter, F. R.: Two-Equation Eddy Viscosity Turbulence Models for Engineering Applications. *AIAA J.* **32**, 1299–1310, 1994.

[15] Launder, B. E. and Sharma, B. I.: Application of the Energy Dissipation Model of Turbulence to the Calculation of Flow Near a Spinning Disc. *Letters in Heat and Mass Transfer* **1**, 131–138, 1974.

[16] Johnson, D. A. and King, L. S.: A Mathematical Simple Turbulence Closure Model for Attached and Separated Turbulent Boundary Layers. *AIAA J.* **23**, 1684–1692, 1985.

[17] Durbin, P. A. and Petterson Reif, B. A.: Statistical Theory and Modeling for Turbulent Flows. John Wiley and Sons, New York, 2001.

2 Solution of the CS and k-ε Model Equations

2.0 Introduction

In this chapter we discuss the numerical solution of the boundary-layer equations using the CS and k-ε models, discussed in Section 1.1 and 1.2 respectively. In Section 2.1 these equations are expressed in transformed variables, which stretch the coordinate normal to the flow and allow large steps to be taken in the x-direction. Section 2.2 discusses the solution procedure for the k-ε model equations with and without wall functions as well as with a zonal method. Numerical solution of the k-ε model equations is addressed in Section 2.3 for a zonal method and in Section 2.4 for model equations with and without wall functions.

2.1 Transformed Variables

We use the Falkner-Skan transformation discussed, for example, in [1]. With the similarity variable defined by

$$\eta = \sqrt{\frac{u_e}{\nu x}}\, y \tag{2.1.1a}$$

and the dimensionless stream function $f(x, \eta)$ by

$$\psi(x, y) = \sqrt{u_e \nu x}\, f(x, \eta), \tag{2.1.1b}$$

the continuity and momentum equations, Eqs. (1.1.1) and (1.1.2) and the k-ε model equations, Eqs. (1.2.18) and (1.2.19), with $\tilde{k}$ and $\tilde{\varepsilon}$ defined by

$$\tilde{k} = \frac{k}{u_e^2}, \qquad \tilde{\varepsilon} = \frac{\varepsilon x}{u_e^3}, \tag{2.1.2}$$

and a prime denoting differentiation with respect to η, can be written as

$$(bf'')' + m_1 f f'' + m[1 - (f')^2] = x \left(f' \frac{\partial f'}{\partial x} - f'' \frac{\partial f}{\partial x} \right) \tag{2.1.3}$$

$$(b_2 k')' + P - Q + F = 2muk - m_1 f k' + x \left(u \frac{\partial k}{\partial x} - k' \frac{\partial f}{\partial x} \right) \tag{2.1.4}$$

$$(b_3 \varepsilon')' + P_1 - Q_1 + E = x \left(u \frac{\partial \varepsilon}{\partial x} - \varepsilon' \frac{\partial f}{\partial x} \right) + (3m - 1)u\varepsilon - m_1 \varepsilon' f \tag{2.1.5}$$

where the tilde has been dropped from the equations and

$$b = 1 + \varepsilon_m^+, \quad b_2 = 1 + \frac{\varepsilon_m^+}{\sigma_k}, \quad b_3 = 1 + \frac{\varepsilon_m^+}{\sigma_\varepsilon}$$
$$\varepsilon_m^+ = \frac{\varepsilon_m}{\nu}, \quad m = \frac{x}{u_e} \frac{du_e}{dx}, \quad m_1 = \frac{m+1}{2} \tag{2.1.6}$$

In Eq. (2.1.4) $(b_2 k')'$ denotes the diffusion term, P and Q defined by

$$P = \varepsilon_m^+ (f'')^2, \quad Q = \varepsilon + D \tag{2.1.7}$$

denote the production and dissipation terms, respectively. The right-hand side of Eq. (2.1.4) represents the convection term.

In Eq. (2.1.5) $(b_3 \varepsilon')'$ denotes the diffusion term, P_1 and Q_1, defined by

$$P_1 = c_{\varepsilon_1} f_1 c_\mu f_\mu v^2 k \tag{2.1.8a}$$

$$Q_1 = c_{\varepsilon_2} f_2 \varepsilon^2 / k \tag{2.1.8b}$$

denote the generation and destruction terms, respectively. The right hand side of Eq. (2.1.5) represents the convection term.

The wall boundary conditions depend on whether the above equations are being solved for high Reynolds flows (without wall functions), for the zonal method or for flows at low Reynolds numbers (with wall functions).

For high Reynolds numbers the four boundary conditions at $y = y_0$ correspond to Eqs. (1.2.6) (1.2.12) and (1.2.14). In terms of transformed variables, Eqs. (1.2.6) can be written as

$$f_0' = w_0 \left[\frac{1}{\kappa} \ln \left(\sqrt{R_x} w_0 \eta_0 \right) + c \right] \tag{2.1.9a}$$

$$x \frac{\partial f_0}{\partial x} + m_1 f_0 = f_0' \eta \left[m_1 + \frac{x}{w_0} \frac{dw_0}{dx} \right] \tag{2.1.9b}$$

where

$$w_0 = \frac{u_\tau}{u_e}, \quad R_x = \frac{u_e x}{\nu}, \quad c = 5.2, \quad \kappa = 0.41 \tag{2.1.10}$$

In terms of dimensionless and transformed variables, Eq. (1.2.12) can be written as

$$a_1 = c_\mu \frac{k}{\varepsilon} \sqrt{R_x} f'' \tag{2.1.11}$$

all evaluated at $\eta = \eta_0$.

The fourth boundary condition, Eq. (1.2.14) becomes

$$\kappa^2 \eta_0^2 f_0'' \left\{ 1 - \exp(-\sqrt{R_x} w_0 \eta_0 / 26) \right\}^2 = c_\mu \sqrt{R_x} \frac{k_0^2}{\varepsilon_0} \tag{2.1.12}$$

For low Reynolds number flows, the wall boundary conditions are given at $\eta = 0$, that is

$$f = 0 \quad \text{(no mass transfer)}, \quad f' = 0 \tag{2.1.13a}$$

$$k = 0, \quad \varepsilon = 0 \tag{2.1.13b}$$

In the zonal method where we use, for example the CS model near the wall region $0 \le y \le y_0$, the wall boundary conditions for f and f' at $\eta = 0$ are given by Eq. (2.1.13a) and those at $\eta = \eta_0$ by Eqs. (2.1.11) and (2.1.12).

In all cases, the "edge" boundary conditions at $\eta = \eta_e$ are given by

$$\eta = \eta_e, \quad u = 1.0, \tag{2.1.14a}$$

$$x \frac{\partial k}{\partial x} + \varepsilon + 2mk = 0, \quad x \frac{\partial \varepsilon}{\partial x} + c_{\varepsilon 2} f_2 \frac{\varepsilon^2}{k} + (3m - 1)\varepsilon = 0 \tag{2.1.14b}$$

2.2 Solution Procedure

The solution procedure of the k-ε model equations can be obtained with and without wall functions. It can also be obtained with a zonal method. In this case the boundary layer equations are solved in two regions with each region employing different turbulence models. In effect this approach may be regarded as the use of the k-ε model with wall functions.

In this book we use Keller's Box method to solve the boundary-layer equations. This is a second order two-point finite-difference method described in detail in several references, see for example [1]. In this method the governing equations are first expressed as a first order system by introducing new functions to represent the derivatives of f, k and ε with respect to η. The first-order equations are approximated on an arbitrary net, Fig. 2.1, with "centered-difference" derivatives and averages at the midpoints of the net rectangle difference equations. The resulting system of equations which is implicit and nonlinear is linearized with Newton's method and solved by the block-elimination discussed in Section 2.4 and also in [1].

With the introduction of new variables

$$f' = u \tag{2.2.1a}$$

$$u' = v \tag{2.2.1b}$$

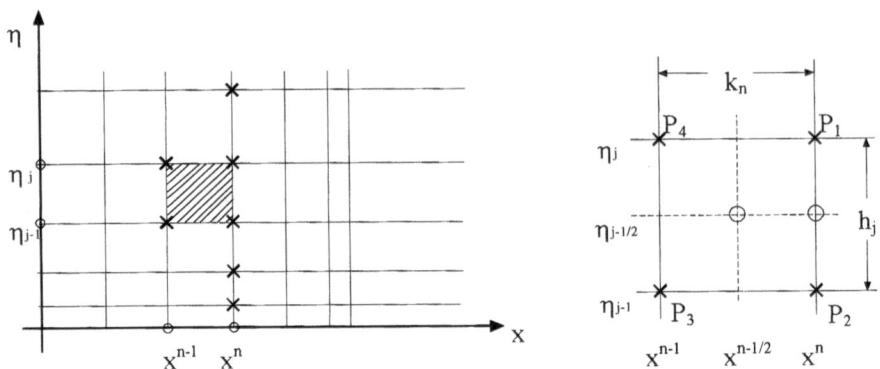

Fig. 2.1. Net rectangle for difference approximations

$$k' = s \tag{2.2.1c}$$

$$\varepsilon' = q , \tag{2.2.1d}$$

the k-ε model equations given by Eqs. (2.1.3) to (2.1.5) can be reduced to a system of seven first-order equations given by Eqs. (2.2.1) and by

$$(b_1 v)' + m_1 f v + m(1 - u^2) = x \left(u \frac{\partial u}{\partial x} - v \frac{\partial f}{\partial x} \right) \tag{2.2.2}$$

$$(b_2 s)' + P - Q + F - 2muk + m_1 fs = \left(u \frac{\partial k}{\partial x} - s \frac{\partial f}{\partial x} \right) \tag{2.2.3}$$

$$(b_3 q)' + P_1 - Q_1 + E + m_1 f q - (3m - 1) u\varepsilon = x \left(u \frac{\partial \varepsilon}{\partial x} - q \frac{\partial f}{\partial x} \right) \tag{2.2.4}$$

As discussed above, the k-ε model equations without wall functions use "wall" boundary conditions specified at some distance η_0 outside the viscous sublayers. In this case the boundary conditions on f' and f (or u and f) are represented by Eqs. (2.1.9), that is

$$u_0' = w_0 \left[\frac{1}{\kappa} \ln \left(\sqrt{R_k w_0 \eta_0} \right) + c \right] \tag{2.2.5a}$$

$$x \frac{\partial f_0}{\partial x} + m_1 f_0 = u\eta \left[m_1 + \frac{x}{w_0} \frac{dw_0}{dx} \right] \tag{2.2.5a}$$

and those for k and ε by Eqs. (2.1.11) and (2.1.12), that is

$$a_1 = c_\mu \frac{k_0}{\varepsilon_0} \sqrt{R_x} v_0 \tag{2.2.6}$$

$$(\varepsilon_m)_{CS} = \kappa^2 \eta_0^2 \left\{ 1 - \exp(-\sqrt{R_x} w_0 \eta_0 / 26) \right\}^2 v_0 = c_\mu \sqrt{R_x} \frac{k_0^2}{\varepsilon_0} \tag{2.2.7}$$

although in the latter case, there are other choices. In either case, the friction velocity, u_τ $(\equiv \sqrt{\frac{\tau_w}{\varrho}})$ appearing in the u and f equations is unknown and must be determined as part of the solution. One approach is to assume u_τ, (say from the initial profiles at the previous x-station), and solve the governing equations subject to the "wall" and edge boundary conditions. From the solution determine τ_0 at y_0,

$$\tau_0 = \varrho(\varepsilon_m)_0 \left(\frac{\partial u}{\partial y}\right)_0 \qquad (2.2.8)$$

and compute u_τ from Eq. (1.2.8). If the calculated value of u_τ does not agree with the estimated value within a specified tolerance parameter δ_1,

$$|u_\tau^{\nu+1} - u_\tau^{\nu}| < \delta_1 \qquad (2.2.9)$$

then a new solution is obtained with the updated values of u_τ and τ_0. This procedure is repeated until convergence.

This iterative procedure can be replaced with a more efficient one by treating u_τ as an unknown. Since u_τ is a function of x only, we can write

$$u_\tau' = 0 \qquad (2.2.10)$$

thus increasing the number of first-order equations from seven to eight. Although we now solve eight first-order equations rather than seven, this procedure allows the solutions to converge faster, especially for flows with strong adverse pressure gradient.

In the solution procedure described here the numerical method is formulated for eight unknowns, not only for the k-ε model equations with the zonal approach but also for the k-ε model equations with and without wall functions. This choice does not increase the complexity of the solution procedure, but paves the way to solve the k-ε model equations or others in an inverse mode if the solution procedure is to be extended to flows with separation (see Section 6.3).

In Keller's Box method, the x-wise derivatives are represented by central differences. Experience with the Box method has shown that when profiles are used to start the turbulent flow calculations, the solutions at the subsequent x-locations oscillate. A common cure to this problem is to compute the first two x-stations equally spaced and take an average of the solutions at the midpoint of x_0 and x_1, say x_m and x_1 and x_2, say x_e. Then another average of the solutions is taken at x_m and x_e defining a new solution at x_1. When new calculations begin at x_2 with averaged profiles at $x = x_1$, the solutions at $x \geq x_2$ do not exhibit oscillations.

While this cure is relatively easy to incorporate into a computer program and in most cases provides stable solutions in adverse pressure gradient flows sometimes the solutions may break down due to oscillations. On the other hand, the author and his colleagues observed that if one uses backward difference approximations for the x-derivatives in the boundary-layer equations, rather than central differences as used in the Box method, the solutions do not oscillate and are more stable. For this reason, when the Box method is used for turbulent flow calculations with initial profiles, we will represent the x-derivatives with backward finite-difference approximations.

2.3 Solution of the k-ε Equations with a Zonal Method

In this method, the boundary-layer is divided into two zones. The inner zone is identified by $y \leq y_0$, $y_0^+ = (y_0 u_\tau / \nu) \approx 100$, where the continuity and momentum equations, Eqs. (1.1.1) and (1.1.3), are solved subject to the wall boundary conditions given by Eq. (1.1.4a), with eddy viscosity ε_m given by the inner region of the CS model. In the outer zone, $y > y_0$, the equations for continuity Eq. (1.1.1), momentum Eq. (1.1.3), turbulence kinetic energy Eq. (1.2.3) and rate of dissipation Eq. (1.2.4) are solved subject to the inner boundary conditions given by Eqs. (1.2.11) and (1.2.14) and the edge boundary conditions given by Eqs. (1.2.15), (1.2.16) and (1.2.17), with eddy viscosity ε_m computed from Eq. (1.0.4).

2.3.1 Inner Region

The numerical solution of the k-ε model equations with the zonal method requires that in the inner region Eq. (2.1.3) is solved subject to the true wall boundary conditions $f = 0$, $u = 0$. Since, however, the solution procedure is being formulated for the general case which includes the solution of the k-ε model equations without wall functions, it is necessary to specify a boundary condition for u_τ. This can be done as described below.

From the definition of u_τ ($\equiv u_e \sqrt{\frac{c_f}{2}}$), we can write

$$\frac{u_\tau}{u_e} \equiv w = \sqrt{\frac{c_f}{2}} \tag{2.3.1a}$$

or in transformed variables,

$$w = \frac{\sqrt{f_w''}}{R_x^{1/4}} . \tag{2.3.1b}$$

The boundary condition for w is

$$w_0 = \frac{\sqrt{v_0}}{R_x^{1/4}} \tag{2.3.2}$$

Next the eight first-order equations can be written by letting $u' = v$, $k' = 0$, $s' = 0$, $\varepsilon' = 0$, $w' = 0$, $q' = 0$, $f' = u$ and the momentum equation (2.2.2). For $j = 0$, with the first three equations corresponding to boundary conditions, *the equations for the inner region are ordered as*

$$f_0 = 0 \tag{2.3.3a}$$

$$u_0 = 0 \tag{2.3.3b}$$

$$w_0 = \frac{\sqrt{v_0}}{R_x^{1/4}} \tag{2.3.3c}$$

$$u' = v \tag{2.3.3d}$$

$$k' = 0 \tag{2.3.3e}$$

$$s' = 0 \tag{2.3.3f}$$

$$\varepsilon' = 0 \tag{2.3.3g}$$

$$q' = 0 \tag{2.3.3h}$$

With finite-difference approximations and linearization, they become

$$\delta f_0 = (r_1)_0 = 0 \tag{2.3.4a}$$

$$\delta u_0 = (r_2)_0 = 0 \tag{2.3.4b}$$

$$\delta v_0 - 2\sqrt{R_x}\, w_0 \delta w_0 = (r_3)_0 = \sqrt{R_x}\, w_0^2 - v_0 \tag{2.3.4c}$$

$$\delta u_j - \delta u_{j-1} - \frac{h_j}{2}(\delta v_j + \delta v_{j-1}) = (r_4)_j = u_{j-1} - u_j + h_j v_{j-1/2} \tag{2.3.4d}$$

$$\delta k_j - \delta k_{j-1} = (r_5)_j = 0 \tag{2.3.4e}$$

$$\delta s_j - \delta s_{j-1} = (r_6)_j = 0 \tag{2.3.4f}$$

$$\delta \varepsilon_j - \delta \varepsilon_{j-1} = (r_7)_j = 0 \tag{2.3.4g}$$

$$\delta q_j - \delta q_{j-1} = (r_8)_j = 0 \tag{2.3.4h}$$

For $1 \leq j \leq j_s$, the order of the equations is the same as those above, except that the first three equations are replaced by

$$w' = 0 \tag{2.3.5a}$$

$$f' = u \tag{2.3.5b}$$

$$\text{momentum Eq. (2.2.2)} \tag{2.3.5c}$$

which, in linearized form can be written as

$$\delta w_j - \delta w_{j-1} = (r_1)_j = 0 \tag{2.3.6a}$$

$$\delta f_j - \delta f_{j-1} - \frac{h_j}{2}(\delta u_j + \delta u_{j-1}) = (r_2)_j = f_{j-1} - f_j + h_j u_{j-1/2} \tag{2.3.6b}$$

$$(s_1)_j \delta f_j + (s_2)_j \delta f_{j-1} + (s_3)_j \delta u_j + (s_4)_j \delta u_{j-1}$$
$$+ (s_5)_j \delta v_j + (s_6)_j \delta v_{j-1} = (r_3)_j \tag{2.3.6c}$$

The finite-difference procedure for Eq. (2.2.2) is identical to the procedure used in the Box method for the momentum equation described in [1]. The only difference occurs in the solution of Eq. (2.2.2) where we use three-point or two-point backward finite-difference formulas for the x-wise derivatives rather than central differences as was done in [1]. For this purpose, for any variable V, the derivative of $\frac{\partial V}{\partial x}$ is defined by

$$\left(\frac{\partial V}{\partial x}\right)^n = A_1 V^{n-2} + A_2 V^{n-1} + A_3 V^n \tag{2.3.7}$$

where for first-order

$$A_1 = 0, \quad A_2 = -\frac{1}{x_n - x_{n-1}}, \quad A_3 = \frac{1}{x_n - x_{n-1}} \tag{2.3.8}$$

and second-order

$$A_1 = \frac{(x_n - x_{n-1})}{(x_{n-2} - x_{n-1})(x_{n-2} - x_n)}$$

$$A_2 = \frac{(x_n - x_{n-2})}{(x_{n-1} - x_{n-2})(x_{n-1} - x_n)} \tag{2.3.9}$$

$$A_3 = \frac{2x_n - x_{n-1} - x_{n-2}}{(x_n - x_{n-2})(x_n - x_{n-1})}$$

Representing the x-derivatives in Eq. (2.2.2) with either *two-point or three-point backward difference approximations at $x = x^n$* and using central differences in the η-direction, we can write Eq. (2.2.2) as

$$h_j^{-1}[(bv)_j^n - (bv)_{j-1}^n] + m_1^n(fv)_{j-1/2}^n + m^n[1 - (u^2)_{j-1/2}^n]$$

$$= \frac{1}{2}x^n\left[\frac{\partial}{\partial x}(u^2)\right]_{j-1/2}^n - \frac{x^n}{2}\left[\left(v\frac{\partial f}{\partial x}\right)_j^n + \left(v\frac{\partial f}{\partial x}\right)_{j-1}^n\right] \tag{2.3.10}$$

Linearizing we get

$$h_j^{-1}(b_j^n \delta v_j - b_{j-1}^n \delta v_{j-1})$$

$$+ \frac{m_1^n}{2}(f_j^n \delta v_j + v_j^n \delta f_j + f_{j-1}^n \delta v_{j-1} + v_{j-1}^n \delta f_{j-1})$$

$$- m^n(u_j \delta u_j + u_{j-1}\delta u_{j-1})$$

$$= \frac{x^n}{4}\left[\frac{\partial}{\partial u}\left(\frac{\partial u^2}{\partial x}\right)_j^n \delta u_j + \frac{\partial}{\partial u}\left(\frac{\partial u^2}{\partial x}\right)_{j-1}^n \delta u_{j-1}\right] \tag{2.3.11}$$

$$- \frac{x^n}{2}\left[\left(\frac{\partial f}{\partial x}\right)_j^n \delta v_j + v_j^n\frac{\partial}{\partial f}\left(\frac{\partial f}{\partial x}\right)_j^n \delta f_j + \left(\frac{\partial f}{\partial x}\right)_{j-1}^n \delta v_{j-1}\right.$$

$$\left. + v_{j-1}^n\frac{\partial}{\partial f}\left(\frac{\partial f}{\partial x}\right)_{j-1}^n \delta f_{j-1}\right] + (r_3)_j$$

From Eq. (2.3.7), it follows that

$$\frac{\partial}{\partial u}\left(\frac{\partial u^2}{\partial x}\right)^n_j = 2A_3 u^n_j, \qquad \frac{\partial}{\partial u}\left(\frac{\partial u^2}{\partial x}\right)^n_{j-1} = 2A_3 u^n_{j-1}$$

$$\frac{\partial}{\partial f}\left(\frac{\partial f}{\partial x}\right)^n_j = A_3, \qquad \frac{\partial}{\partial f}\left(\frac{\partial f}{\partial x}\right)^n_{j-1} = A_3 \tag{2.3.12}$$

The linearized expression can be written in the form given by Eq. (2.3.6c). The coefficients $(s_1)_j$ to $(s_6)_j$ and $(r_3)_j$ are given by

$$(s_1)_j = \frac{1}{2}(m^n_1 + x^n A_3)v^n_j \tag{2.3.12a}$$

$$(s_2)_j = \frac{1}{2}(m^n_1 + x^n A_3)v^n_{j-1} \tag{2.3.12b}$$

$$(s_3)_j = -\left(m^n + \frac{x^n}{2}A_3\right)u^n_j \tag{2.3.12c}$$

$$(s_4)_j = -\left(m^n + \frac{x^n}{2}A_3\right)u^n_{j-1} \tag{2.3.12d}$$

$$(s_5)_j = h^{-1}_j b^n_j + \frac{m^n_1}{2}f^n_j + \frac{x^n}{2}\left(\frac{\partial f}{\partial x}\right)^n_j \tag{2.3.13e}$$

$$(s_6)_j = -h^{-1}_j b^n_{j-1} + \frac{m^n_1}{2}f^n_{j-1} + \frac{x^n}{2}\left(\frac{\partial f}{\partial x}\right)_{j-1} \tag{2.3.13f}$$

$$(r_3)_j = -\left[h^{-1}_j[(bv)^n_j - (bv)^n_{j-1}] + m^n_1(fv)^n_{j-1/2} + m^n[1 - (u^2)^n_{j-1/2}]\right]$$

$$+ \frac{1}{2}x^n\left[\frac{\partial}{\partial x}(u^2)\right]^n_{j-1/2} - \frac{x^n}{2}\left[\left(v\frac{\partial f}{\partial x}\right)^n_j + \left(v\frac{\partial f}{\partial x}\right)^n_{j-1}\right] \tag{2.3.14}$$

The linearized finite-difference equations and their boundary conditions, Eqs. (2.3.4) and (2.3.6) are written in matrix-vector form as described in [1]

$$A\vec{\delta} = \vec{r} \tag{2.3.15}$$

where

$$A = \begin{vmatrix} A_0 & C_0 & & & & \\ B_1 & A_1 & C_1 & & & \\ & \cdot & \cdot & \cdot & & \\ & & B_j & A_j & C_j & \\ & & & \cdot & \cdot & \cdot \\ & & & B_{J-1} & A_{J-1} & C_{J-1} \\ & & & & B_J & A_J \end{vmatrix} \quad \vec{\delta} = \begin{vmatrix} \vec{\delta}_0 \\ \vec{\delta}_1 \\ \cdot \\ \cdot \\ \vec{\delta}_j \\ \cdot \\ \cdot \\ \vec{\delta}_J \end{vmatrix} \quad \vec{r} = \begin{vmatrix} \vec{r}_0 \\ \vec{r}_1 \\ \cdot \\ \cdot \\ \vec{r}_j \\ \cdot \\ \cdot \\ \vec{r}_J \end{vmatrix} \tag{2.3.16}$$

The eight dimensional vectors $\vec{\delta}_j$ and $\vec{r}_j$ for each value of j are defined by

$$
\vec{\delta}_j = \begin{vmatrix} \delta f_j \\ \delta u_j \\ \delta v_j \\ \delta k_j \\ \delta s_j \\ \delta \varepsilon_j \\ \delta q_j \\ \delta w_j \end{vmatrix}, \quad
\vec{r}_j = \begin{vmatrix} (r_1)_j \\ (r_2)_j \\ (r_3)_j \\ (r_4)_j \\ (r_5)_j \\ (r_6)_j \\ (r_7)_j \\ (r_8)_j \end{vmatrix}
\tag{2.3.17}
$$

The definitions of 8×8 matrices A_j, B_j and C_j in the inner region $0 \le j \le j_s$ are

$$
A_0 = \begin{vmatrix}
1 & 0 & 0 & 0 & 0 & 0 & 0 & 0 \\
0 & 1 & 0 & 0 & 0 & 0 & 0 & 0 \\
0 & 0 & 1 & 0 & 0 & 0 & 0 & -2\sqrt{R_x}\,w_0 \\
0 & -1 & -\frac{h_1}{2} & 0 & 0 & 0 & 0 & 0 \\
0 & 0 & 0 & -1 & 0 & 0 & 0 & 0 \\
0 & 0 & 0 & 0 & -1 & 0 & 0 & 0 \\
0 & 0 & 0 & 0 & 0 & -1 & 0 & 0 \\
0 & 0 & 0 & 0 & 0 & 0 & -1 & 0
\end{vmatrix}
\tag{2.3.18a}
$$

$$
C_j = \begin{vmatrix}
0 & 0 & 0 & 0 & 0 & 0 & 0 & 0 \\
0 & 0 & 0 & 0 & 0 & 0 & 0 & 0 \\
0 & 0 & 0 & 0 & 0 & 0 & 0 & 0 \\
0 & 1 & -\frac{h_{j+1}}{2} & 0 & 0 & 0 & 0 & 0 \\
0 & 0 & 0 & 1 & 0 & 0 & 0 & 0 \\
0 & 0 & 0 & 0 & 1 & 0 & 0 & 0 \\
0 & 0 & 0 & 0 & 0 & 1 & 0 & 0 \\
0 & 0 & 0 & 0 & 0 & 0 & 1 & 0
\end{vmatrix} \quad 0 \le j \le j_s - 1
\tag{2.3.18b}
$$

$$
A_j = \begin{vmatrix}
0 & 0 & 0 & 0 & 0 & 0 & 0 & 1 \\
1 & -\frac{h_j}{2} & 0 & 0 & 0 & 0 & 0 & 0 \\
(s_1)_j & (s_3)_j & (s_5)_j & 0 & 0 & 0 & 0 & 0 \\
0 & -1 & -\frac{h_{j+1}}{2} & 0 & 0 & 0 & 0 & 0 \\
0 & 0 & 0 & -1 & 0 & 0 & 0 & 0 \\
0 & 0 & 0 & 0 & -1 & 0 & 0 & 0 \\
0 & 0 & 0 & 0 & 0 & -1 & 0 & 0 \\
0 & 0 & 0 & 0 & 0 & 0 & -1 & 0
\end{vmatrix} \quad 1 \le j \le j_s - 1
\tag{2.3.18c}
$$

$$B_j = \begin{vmatrix} 0 & 0 & 0 & 0 & 0 & 0 & 0 & -1 \\ -1 & -\frac{h_j}{2} & 0 & 0 & 0 & 0 & 0 & 0 \\ (s_2)_j & (s_4)_j & (s_6)_j & 0 & 0 & 0 & 0 & 0 \\ 0 & 0 & 0 & 0 & 0 & 0 & 0 & 0 \\ 0 & 0 & 0 & 0 & 0 & 0 & 0 & 0 \\ 0 & 0 & 0 & 0 & 0 & 0 & 0 & 0 \\ 0 & 0 & 0 & 0 & 0 & 0 & 0 & 0 \\ 0 & 0 & 0 & 0 & 0 & 0 & 0 & 0 \end{vmatrix} \quad 1 < j \le j_s \qquad (2.3.18d)$$

2.3.2 Interface Between Inner and Outer Regions

The first-order system of equations is now ordered as

$$w' = 0 \qquad (2.3.19a)$$

$$f' = u \qquad (2.3.19b)$$

$$\text{momentum Eq. (2.2.2)} \qquad (2.3.19c)$$

$$\text{b.c. Eq. (2.2.7)} \qquad (2.3.19d)$$

$$\text{b.c. Eq. (2.2.6)} \qquad (2.3.19e)$$

$$u' = v \qquad (2.3.19f)$$

$$k' = s \qquad (2.3.19g)$$

$$\varepsilon' = q \qquad (2.3.19h)$$

The resulting A_j and C_j matrices from the linearized equations, with B_j given by Eq. (2.3.18d) and $(s_1)_j$ to $(s_6)_j$ by Eq. (2.3.13) at $j = j_s$ are

$$A_{j_s} = \begin{vmatrix} 0 & 0 & 0 & 0 & 0 & 0 & 0 & 1 \\ 1 & -\frac{h_{j_s}}{2} & 0 & 0 & 0 & 0 & 0 & 0 \\ (s_1)_{j_s} & (s_3)_{j_s} & (s_5)_{j_s} & 0 & 0 & 0 & 0 & 0 \\ 0 & 0 & D_1 & D_2 & 0 & D_3 & 0 & 0 \\ 0 & 0 & D_4 & D_5 & 0 & D_6 & 0 & 0 \\ 0 & -1 & -\frac{h_{j_s+1}}{2} & 0 & 0 & 0 & 0 & 0 \\ 0 & 0 & 0 & -1 & -\frac{h_{j_s+1}}{2} & 0 & 0 & 0 \\ 0 & 0 & 0 & 0 & 0 & -1 & -\frac{h_{j_s+1}}{2} & 0 \end{vmatrix} \qquad (2.3.20a)$$

$$C_{j_s} = \begin{vmatrix} 0 & 0 & 0 & 0 & 0 & 0 & 0 & 0 \\ 0 & 0 & 0 & 0 & 0 & 0 & 0 & 0 \\ 0 & 0 & 0 & 0 & 0 & 0 & 0 & 0 \\ 0 & 0 & 0 & 0 & 0 & 0 & 0 & 0 \\ 0 & 0 & 0 & 0 & 0 & 0 & 0 & 0 \\ 0 & 1 & -\frac{h_{j_s+1}}{2} & 0 & 0 & 0 & 0 & 0 \\ 0 & 0 & 0 & 1 & -\frac{h_{j_s+1}}{2} & 0 & 0 & 0 \\ 0 & 0 & 0 & 0 & 0 & 1 & -\frac{h_{j_s+1}}{2} & 0 \end{vmatrix} \qquad (2.3.20b)$$

Here the fourth and fifth rows of A_{j_s} follow from the boundary conditions, Eqs. (2.2.6) and (2.2.7), at $\eta = \eta_0$. After the application of Newton's method to these equations, D_1 to D_6 are given by the following expressions.

$$D_1 = \varepsilon_{j_s} \frac{\partial}{\partial v}(\varepsilon_m^+)_{\mathrm{CS}} , \quad D_2 = -2\sqrt{R_x}\, c_\mu k_{j_s} , \quad D_3 = (\varepsilon_m^+)_{\mathrm{CS}} \tag{2.3.21a}$$

$$D_4 = 2R_x c_\mu k_{j_s}^2 v_{j_s}, \quad D_5 = 2R_x c_\mu v_{j_s}^2 k_{j_s}, \quad D_6 = -2\varepsilon_{j_s} \tag{2.3.21b}$$

The associated $(r_4)_{j_s}$ and $(r_5)_{j_s}$ are

$$(r_4)_{j_s} = \sqrt{R_x}\, c_\mu k_{j_s}^2 - (\varepsilon_m^+)_{\mathrm{CS}} \varepsilon_{j_s} \tag{2.3.22a}$$

$$(r_5)_{j_s} = \varepsilon_{j_s}^2 - R_x c_\mu (k_{j_s} v_{j_s})^2 \tag{2.3.22b}$$

2.3.3 Outer Region

The finite-difference approximations for the outer region defined for $j_s + 1 \le j \le J$ are written by using a similar procedure to that described for the inner region equations. The first-order system of equations is ordered similar to those given by Eqs. (2.3.19) except that Eqs. (2.3.19d) and (2.3.19e) are replaced by Eqs. (2.2.3) and (2.2.4). The resulting matrices from the linearized equations, with the C_j matrix remaining the same as that given by Eq. (2.3.20b) for $j_s < j \le J - 1$, are

$$B_j = \begin{vmatrix} 0 & 0 & 0 & 0 & 0 & 0 & 0 & -1 \\ -1 & -\frac{h_j}{2} & 0 & 0 & 0 & 0 & 0 & 0 \\ (s_2)_j & (s_4)_j & (s_6)_j & (s_8)_j & 0 & (s_{12})_j & 0 & 0 \\ (\alpha_2)_j & (\alpha_4)_j & (\alpha_6)_j & (\alpha_8)_j & (\alpha_{10})_j & (\alpha_{12})_j & 0 & 0 \\ (\beta_2)_j & (\beta_4)_j & (\beta_6)_j & (\beta_8)_j & 0 & (\beta_{12})_j & (\beta_{14})_j & 0 \\ 0 & 0 & 0 & 0 & 0 & 0 & 0 & 0 \\ 0 & 0 & 0 & 0 & 0 & 0 & 0 & 0 \\ 0 & 0 & 0 & 0 & 0 & 0 & 0 & 0 \end{vmatrix} \quad j_s + 1 < j \le J$$

$$\tag{2.3.23a}$$

$$A_j = \begin{vmatrix} 0 & 0 & 0 & 0 & 0 & 0 & 0 & 1 \\ 1 & -\frac{h_j}{2} & 0 & 0 & 0 & 0 & 0 & 0 \\ (s_1)_j & (s_3)_j & (s_5)_j & (s_7)_j & 0 & (s_{11})_j & 0 & 0 \\ (\alpha_1)_j & (\alpha_3)_j & (\alpha_5)_j & (\alpha_7)_j & (\alpha_9)_j & (\alpha_{11})_j & 0 & 0 \\ (\beta_1)_j & (\beta_3)_j & (\beta_5)_j & (\beta_7)_j & 0 & (\beta_{11})_j & (\beta_{13})_j & 0 \\ 0 & -1 & -\frac{h_{j+1}}{2} & 0 & 0 & 0 & 0 & 0 \\ 0 & 0 & 0 & -1 & -\frac{h_{j+1}}{2} & 0 & 0 & 0 \\ 0 & 0 & 0 & 0 & 0 & -1 & -\frac{h_{j+1}}{2} & 0 \end{vmatrix} \quad j_s + 1 < j \le J - 1$$

$$\tag{2.3.23b}$$

$$A_J = \begin{vmatrix} 0 & 0 & 0 & 0 & 0 & 0 & 0 & 1 \\ 1 & -\frac{h_J}{2} & 0 & 0 & 0 & 0 & 0 & 0 \\ (s_1)_J & (s_3)_J & (s_5)_J & (s_7)_J & 0 & (s_{11})_J & 0 & 0 \\ (\alpha_1)_J & (\alpha_3)_J & (\alpha_5)_J & (\alpha_7)_J & (\alpha_9)_J & (\alpha_{11})_J & 0 & 0 \\ (\beta_1)_J & (\beta_3)_J & (\beta_5)_J & (\beta_7)_J & 0 & (\beta_{11})_J & (\beta_{13})_J & 0 \\ 0 & 1 & 0 & 0 & 0 & 0 & 0 & 0 \\ 0 & 0 & 0 & E_1 & 0 & E_2 & 0 & 0 \\ 0 & 0 & 0 & E_3 & 0 & E_4 & 0 & 0 \end{vmatrix} \tag{2.3.23c}$$

Here $(s_1)_j$ to $(s_{12})_j$, $(\alpha_1)_j$, to $(\alpha_{12})_j$ and $(\beta_1)_j$ to $(\beta_{14})_j$ given in subsection 3.3.1 correspond to the coefficients of the linearized momentum (2.2.2), kinetic energy of turbulence (2.2.3), and rate of dissipation (2.2.4) equations written in the following forms, respectively,

$$(s_1)_j \delta f_j + (s_2)_j \delta f_{j-1} + (s_3)_j \delta u_j + (s_4)_j \delta u_{j-1} + (s_5)_j \delta v_j$$
$$+ (s_6)_j \delta v_{j-1} + (s_7)_j \delta k_j + (s_8)_j \delta k_{j-1} + (s_{11})_j \delta \varepsilon_j$$
$$+ (s_{12})_j \delta \varepsilon_{j-1} = (r_3)_j \tag{2.3.24}$$

$$(\alpha_1)_j \delta f_j + (\alpha_2)_j \delta f_{j-1} + (\alpha_3)_j \delta u_j + (\alpha_4)_j \delta u_{j-1} + (\alpha_5)_j \delta v_j$$
$$+ (\alpha_6)_j \delta v_{j-1} + (\alpha_7)_j \delta k_j + (\alpha_8)_j \delta k_{j-1} + (\alpha_9)_j \delta s_j$$
$$+ (\alpha_{10})_j \delta s_{j-1} + (\alpha_{11})_j \delta \varepsilon_j + (\alpha_{12})_j \delta \varepsilon_{j-1} = (r_4)_j \tag{2.3.25}$$

$$(\beta_1)_j \delta f_j + (\beta_2)_j \delta f_{j-1} + (\beta_3)_j \delta u_j + (\beta_4)_j \delta u_{j-1} + (\beta_5)_j \delta v_j$$
$$+ (\beta_6)_j \delta v_{j-1} + (\beta_7)_j \delta k_j + (\beta_8)_j \delta k_{j-1} + (\beta_{11})_j \delta \varepsilon_j$$
$$+ (\beta_{12})_j \delta \varepsilon_{j-1} + (\beta_{13})_j \delta q_j + (\beta_{14})_j \delta q_{j-1} = (r_5)_j \tag{2.3.26}$$

The last three rows of the A_J matrix correspond to the edge boundary conditions and follow from the linearized forms of Eq. (2.1.14). They are given by

$$E_1 = 2m^n + x^n \frac{\partial}{\partial k}\left(\frac{\partial k}{\partial x}\right)_J^n,$$

$$E_2 = 1, \qquad E_3 = -c_{\varepsilon 2} f_2^n \frac{(\varepsilon^2)_J^n}{(k^2)_J^n}, \tag{2.3.27}$$

$$E_4 = 3m^n - 1 + x^n \frac{\partial}{\partial \varepsilon}\left(\frac{\partial \varepsilon}{\partial x}\right)_J^n + c_{\varepsilon 2} f_2^n \frac{2\varepsilon_J^n}{k_J^n}$$

where

$$\frac{\partial}{\partial k}\left(\frac{\partial k}{\partial x}\right)_J^n = A_3, \qquad \frac{\partial}{\partial \varepsilon}\left(\frac{\partial \varepsilon}{\partial x}\right)_J^n = A_3 \tag{2.3.28}$$

The coefficients $(r_7)_J$ and $(r_8)_J$ are given by

$$(r_7)_J = -\left[x^n \left(\frac{\partial k}{\partial x}\right)_J^n + \varepsilon_J^n + 2m^n k_J^n\right] \tag{2.3.29a}$$

$$(r_8)_J = -\left[x^n \left(\frac{\partial \varepsilon}{\partial x}\right)_J^n + c_{\varepsilon 2} f_2^n \frac{(\varepsilon^2)_J^n}{k_J^n} + (3m^n - 1)\varepsilon_J^n\right] \tag{2.3.29b}$$

2.3.4 Block-Elimination Method

The linear system expressed in the form of Eq. (2.3.15) can be solved by the block-elimination method discussed by Cebeci and Cousteix [1]. According to this method, the solution procedure consists of two sweeps. In the first part of the so-called *forward* sweep, we compute Γ_j, Δ_j from the recursion formulas given by

$$\Delta_0 = A_0 \tag{2.3.30a}$$

$$\Gamma_j \Delta_{j-1} = B_j \quad j = 1, 2, \ldots, J \tag{2.3.30b}$$

$$\Delta_j = A_j - \Gamma_j C_{j-1} \quad j = 1, 2, \ldots, J \tag{2.3.30c}$$

where the Γ_j matrix has the same structure as B_j. In the second part of the forward sweep, we compute $\tilde{w}_j$ from the following relations

$$\tilde{w}_0 = \tilde{r}_0 \tag{2.3.31a}$$

$$\tilde{w}_j = \tilde{r}_j - \Gamma_j \tilde{w}_{j-1} \quad 1 \leq j \leq J \tag{2.3.31b}$$

In the so-called *backward* sweep, we compute $\vec{\delta}_j$ from the recursion formulas given by

$$\Delta_J \vec{\delta}_J = \vec{w}_J \tag{2.3.32a}$$

$$\Delta_j \vec{\delta}_j = \vec{w}_j - C_j \vec{\delta}_{j+1} \quad j = J-1, J-2, \ldots, 0 \tag{2.3.32b}$$

The block elimination method is a general one and can be used to solve any system of first-order equations. The amount of algebra in solving the recursion formulas given by Eqs. (2.3.30) to (2.3.32), however, depends on the order of the matrices A_j, B_j, C_j. When it is small, the matrices Γ_j, Δ_j and the vector $\vec{w}_j$ can be obtained by relatively simple expressions, as discussed in [1]. However, this procedure, though very efficient, becomes increasingly tedious as the order of matrices increases and requires the use of an algorithm that reduces the algebra internally. A general algorithm, called the "matrix solver", discussed by Cebeci and Cousteix [1] and in Section 3.4 can be used for this purpose.

In addition, since the zonal method requires that the linearized inner boundary conditions resulting from Eqs. (2.2.6) and (2.2.7) be satisfied, as well as the usual boundary conditions at the surface and the boundary-layer edge. Subsection 3.3.5 presents an algorithm utilizing the "matrix solver" and called KE-SOLV for this purpose. It employs the block-elimination method and follows the structure of the solution procedure used in the zonal method, as well as the procedure used in the solution of the k-ε model equations with and without wall functions discussed in the following section.

2.4 Solution of the k-ε Model Equations with and Without Wall Functions

The solution of the k-ε model equations with and without wall functions is similar to the solution of the k-ε model equations with the zonal method. Their solution in either case can be accomplished with minor changes to the solution algorithm described in the previous section. In both cases changes are made to the A_{j_s} matrix, Eq. (2.3.20a), by modifying or redefining the elements of the first five rows which in this case correspond to the boundary conditions at $\eta = \eta_0$ or $\eta = 0$. In either case, for $j = 0$, after the five boundary conditions are specified, the next three equations correspond to those given by Eqs. (2.3.19f) to (2.3.19h). For $j \geq 1$, the ordering of the first-order equations is identical to that used for the outer region, that is, the equations are ordered according to those given by Eqs. (2.3.19) except that Eqs. (2.3.19d) and (2.3.19e) are replaced by Eqs. (2.2.3) and (2.2.4), respectively. In addition of course, the coefficients of the equations for the momentum, kinetic energy and rate of dissipation are different.

2.4.1 Solution of the k-ε Model Equations Without Wall Functions

The transformed k-ε model equations without wall functions for high Reynolds number flows are still given by Eqs. (2.2.3) and (2.2.4) provided we set

$$P = \varepsilon_m^+ v^2, \quad Q = \varepsilon, \quad F = 0$$

$$P_1 = c_{\varepsilon_1} f_1 c_\mu v^2 k, \quad Q_1 = c_{\varepsilon_2} f_2 \frac{\varepsilon^2}{k}, \quad E = 0, \quad f_1 = f_2 = 1.0.$$

(2.4.1)

There are five "wall" boundary conditions; four are given by Eqs. (2.2.5), (2.2.6) and (2.2.7) for $\eta = \eta_0$. After linearization they can be expressed in the form

$$\delta u_0 + \alpha_8 \delta w_0 = (r_1)_0 \tag{2.4.2}$$

$$\beta_1 \delta f_0 + \beta_2 \delta u_0 + \beta_8 \delta w_0 = (r_2)_0 \tag{2.4.3}$$

$$\gamma_3 \delta v_0 + \gamma_4 \delta k_0 + \gamma_6 \delta \varepsilon = (r_3)_0 \tag{2.4.4}$$

$$\theta_3 \delta v_0 + \theta_4 \delta k_0 + \theta_6 \delta \varepsilon_0 = (r_4)_0 \tag{2.4.5}$$

where

$$\alpha_8 = -\frac{1}{\kappa} - \left[\frac{1}{\kappa} \ln(\sqrt{R_x}\, w_0 \eta_0) + c \right], \quad \beta_1 = \alpha + m_1, \quad \alpha = \frac{x_n}{k_n} \tag{2.4.6a}$$

$$\beta_2 = -\eta_0 \left[m_1 + \alpha \left(1 - \frac{w_0^{n-1}}{w_0^n} \right) \right], \quad \beta_8 = -u_0 \eta_0 \alpha \frac{w_0^{n-1}}{(w_0^n)^2} \tag{2.4.6b}$$

$$\gamma_3 = c_\mu \sqrt{R_x}\, k_0, \quad \gamma_4 = c_\mu \sqrt{R_x}\, v_0, \quad \gamma_6 = -\sqrt{c_\mu} \tag{2.4.6c}$$

$$\theta_3 = -c_v \varepsilon_0 , \quad \theta_4 = 2c_\mu \sqrt{R_x}\, k_0 , \quad \theta_6 = -c_v v_0 , \quad c_v = (\kappa \eta \cdot \text{damping})^2 \tag{2.4.6d}$$

$$(r_1)_0 = w_0 \left[\frac{1}{\kappa} \ln(\sqrt{R_x}\, w_0 \eta_0) + c \right] - u_0 \tag{2.4.7a}$$

$$(r_2)_0 = u_0 \eta_0 \left[m_1 + \alpha \left(1 - \frac{w_0^{n-1}}{w_0^n} \right) \right] - \alpha(f_0^n - f_0^{n-1}) - m_1 f_0^n \tag{2.4.7b}$$

$$r_3 = \sqrt{c_\mu}\, \varepsilon_0 - c_\mu \sqrt{R_x}\, k_0 v_0 . \tag{2.4.7c}$$

$$r_4 = c_v v_0 \varepsilon_0 - c_\mu \sqrt{R_x}\, k_0^2 \tag{2.4.7d}$$

If η_0 is sufficiently away from the wall, i.e. $y_0^+ \geq 60$, then the damping term, such as the one used in the CS model, is equal to 1.0.

The fifth "wall" boundary condition which connects τ_0 at $y = y_0$ and τ_w at $y = 0$, is obtained from Eq. (1.2.8c). With Thompson's and log law velocity profiles, it can be written as

$$\tau_0 = \tau_w + \alpha^* y_0 \frac{d\tau_w}{dx} + y_0 \frac{dp}{dx} \tag{2.4.8}$$

Here α^* is given by

$$\alpha^* = 0.5 \left[c_1 (\ln y_0^+)^2 + c_2 \ln y_0^+ + c_3 + \frac{c_4}{y_0^+} \right] \tag{2.4.9}$$

where

$$c_1 = 5.9488 , \quad c_2 = 13.4682 , \quad c_3 = 13.5718 , \quad c_4 = -785.20$$
$$y_0^+ = \sqrt{R_x}\, \bar{u}_\tau \eta_0 , \quad y_0 = \frac{x \eta_0}{\sqrt{R_x}} \tag{2.4.10}$$

In terms of transformed variables, Eq. (2.4.8), after linearization, can be expressed in the form

$$\delta_3 \delta v_0 + \delta_4 \delta k_0 + \delta_6 \delta \varepsilon_0 + \delta_8 \delta w_0 = (r_5)_0 \tag{2.4.11}$$

where

$$\delta_3 = c_\mu \sqrt{R_x}\, \frac{k_0^2}{\varepsilon_0} , \quad \delta_4 = 2c_\mu \sqrt{R_x}\, \frac{k_0}{\varepsilon_0} v_0 , \quad \delta_6 = -c_\mu \sqrt{R_x}\, \frac{k_0^2}{\varepsilon_0^2} v_0 \tag{2.4.12a}$$

$$\delta_8 = -2w_0 \{ \sqrt{R_x} + \alpha^* \eta_0 (\alpha + 2m_1) \} -$$
$$- \left(\frac{\partial \alpha^*}{\partial w} \right)_0 \eta_0 \{ \alpha[(w_0^n)^2 - (w_0^{n-1})^2] + 2m_1 (w_0^n)^2 \} \tag{2.4.12b}$$

and

$$(r_5)_0 = \sqrt{R_x}\, w^2 + \alpha^* \eta_0 \{\alpha[(w_0^n)^2 - (w_0^{n-1})^2] + 2m_1(w_0^n)^2\}$$

$$- \eta_0 m_1 - c_\mu \sqrt{R_x} \frac{k_0^2}{\varepsilon_0} v_0 \tag{2.4.13}$$

With the five boundary conditions defined, the A_{j_s} matrix, which is essentially the A_0 matrix in this case, becomes

$$A_0 = \begin{vmatrix} 0 & 1 & 0 & 0 & 0 & 0 & 0 & \alpha_8 \\ \beta_1 & \beta_2 & 0 & 0 & 0 & 0 & 0 & \beta_8 \\ 0 & 0 & \gamma_3 & \gamma_4 & 0 & \gamma_6 & 0 & 0 \\ 0 & 0 & \theta_3 & \theta_4 & 0 & \theta_6 & 0 & 0 \\ 0 & 0 & \delta_3 & \delta_4 & 0 & \delta_6 & 0 & \delta_8 \\ 0 & -1 & -\frac{h_1}{2} & 0 & 0 & 0 & 0 & 0 \\ 0 & 0 & 0 & -1 & -\frac{h_1}{2} & 0 & 0 & 0 \\ 0 & 0 & 0 & 0 & 0 & -1 & -\frac{h_1}{2} & 0 \end{vmatrix} \tag{2.4.14}$$

2.4.2 Solution of the k-ε Model Equations with Wall Functions

The solution of the k-ε model equations with wall functions is similar to the procedure described for the case without wall functions. Again the only changes occur in the first five rows of the A_{j_s} matrix, Eq. (2.3.20a). Of the five boundary conditions at the wall, the first three are written in the order given by Eqs. (2.3.3a,b,c) and the fourth and fifth are given by

$$k_0 = 0 \tag{2.4.15a}$$
$$\varepsilon_0 = 0 \tag{2.4.15b}$$

or in linearized form

$$\delta k_0 = (r_4)_0 = 0 \tag{2.4.16a}$$
$$\delta \varepsilon_0 = (r_5)_0 = 0 \tag{2.4.16b}$$

The structure of the other matrices remains the same but, of course, the coefficients of the linearized momentum, kinetic energy and dissipation equations, Eqs. (2.3.24) to (2.3.26), respectively are different than those for k-ε model equations without wall functions. These coefficients naturally vary depending on the wall functions used.

The A_0 matrix for the k-ε model equations with wall functions, with the last three rows identical to those in Eq. (2.4.14), is

$$A_0 = \begin{vmatrix} 1 & 0 & 0 & 0 & 0 & 0 & 0 & 0 \\ 0 & 1 & 0 & 0 & 0 & 0 & 0 & 0 \\ 0 & 0 & 1 & 0 & 0 & 0 & 0 & -2\sqrt{R_x}\, w_0 \\ 0 & 0 & 0 & 1 & 0 & 0 & 0 & 0 \\ 0 & 0 & 0 & 0 & 0 & 1 & 0 & 0 \\ 0 & -1 & -\frac{h_1}{2} & 0 & 0 & 0 & 0 & 0 \\ 0 & 0 & 0 & -1 & -\frac{h_1}{2} & 0 & 0 & 0 \\ 0 & 0 & 0 & 0 & 0 & -1 & -\frac{h_1}{2} & 0 \end{vmatrix} \tag{2.4.17}$$

References

[1] Cebeci, T. and Cousteix, J.: Modeling and Computation of Boundary-Layer Flows, Horizons Publishing, Long Beach, CA and Springer-Verlag, Heidelberg, Germany, 1998.

3 Computer Program for CS and k-ε Models

3.0 Introduction

In this chapter we present and discuss a code for two-dimensional incompressible turbulent flows. The code, given in the accompanying CD-ROM, includes a computer program employing the CS model discussed in Section 1.1 and another program employing the k-ε model discussed in Section 1.2. It is prepared in five parts. In Part 1, discussed in Section 3.1, the components of the computer program common to both models are described. Part 2, discussed in Section 3.2, describes the components of the CS model, and Parts 3 and 4 describe the k-ε model and its solution algorithm in Sections 3.3 and 3.4, respectively. The computer program for the k-ε model includes the zonal method with a combination of the CS model for the inner region and the k-ε model for the outer region as discussed in Section 2.3. It also includes the solution of the k-ε model equations with and without wall functions. Finally Part 5 includes a brief description of the basic tools such as integration, smoothing and differentiation, used in the computer program.

3.1 Part 1 – Components of the Computer Program Common to Both Models

Part 1 includes a MAIN routine which contains the logic of the computations and five subroutines: INPUT, IVPT, GROWTH, GRID and OUTPUT. The following subsections present a brief description of each routine.

3.1.1 MAIN

Here we first read in input data (subroutine INPUT) and generate the initial turbulent velocity profile (subroutine IVPT) and the eddy viscosity distribution for the CS model (subroutine EDDY), k-profile (subroutine KEINITK), ε-profile (subroutine KEINITK). Since linearized equations are being solved, we use an iteration procedure in which the solutions of the equations are obtained for successive estimates of velocity, kinetic energy, dissipation profiles with a subsequent need to check the convergence of the solutions. A convergence criterion based on $\frac{\delta v_0}{v_0}$ is used and the iterations are stopped when

$$\left| \frac{\delta v_0}{v_0} \right| < 0.02$$

During this iteration procedure, we introduce an under-relaxation procedure for the iterations as described in MAIN. This is useful, especially with transport equation turbulence models.

When the solutions converge, we also check to see whether the boundary-layer thickness, η_e, used in the calculations for that x-station is large enough so that the asymptotic behavior of the solutions is reached. If this is not the case, we call subroutine GROWTH.

After the convergence of the solutions, the OUTPUT subroutine is called and the profiles which represent the variables such as f_j, u_j, v_j, k_j, ε_j etc. are shifted.

3.1.2 Subroutine INPUT

In this subroutine we read in input data and set up the flow calculations according to the following turbulence models listed below.

$$
\begin{aligned}
\text{Model} = 0 & \quad \text{CS model} \\
1 & \quad \text{Huang-Lin } k\text{-}\varepsilon \text{ model} \\
2 & \quad \text{Chien } k\text{-}\varepsilon \text{ model} \\
= -1 & \quad \text{zonal method} \\
= -2 & \quad \text{high Re } \# \ k\text{-}\varepsilon \text{ model}
\end{aligned}
$$

In some problems, like airfoil flows, it is convenient to read in the dimensionless airfoil coordinates x/c, y/c rather than the surface distance required in the boundary-layer calculations. In all calculations, the external velocity $u_e(x)$ either dimensional or dimensionless, u_e/u_∞, and freestream or reference velocity, $u_\infty(u_{\mathrm{ref}})$, kinematic viscosity ν (CNU), reference length c (chord), variable η-grid parameter K (VGP) discussed in subroutine GRID must be specified together with R_θ (RTHA) and c_f (CFA) needed to generate the initial turbulent velocity profile with subroutine IVPT.

The input also requires the specification of the first grid point needed in the η-grid generated by subroutine GRID. This is done by inputting y_0^+ (YPLUSW). defined by

$$\frac{y_0^+ u_\tau}{\nu}$$

where u_τ (UTAU) is the friction velocity, $u_e \sqrt{c_f/2}$ and y_0 is the variable grid parameter h_1, discussed in subroutine GRID. Its typical values for CS, zonal and high Reynolds number k-ε models are around 0.5 to 1.0. For low Reynolds number k-ε model, values of y_0^+ around 0.10 to 0.50 are typical. In the present program, K is set equal to 1.12, y_0^+ equal to 0.5 for low Reynolds number k-ε model and 1.0 for zonal and CS models.

Since equations use transformed variables where y is given by

$$y = \sqrt{\nu x / u_e} \eta$$

and since the location of x where the turbulent flow calculations are started, x_1, can be an arbitrary distance, in this subroutine we calculate x_0 in order to control y_0^+ better.

The calculation of the pressure gradient parameter $m(x)$ (P2) in the transformed momentum equation is achieved from the given external velocity $u_e(x)$ distribution and from the definition of m. The derivative of du_e/dx is obtained by using subroutine DIFF3 given in Section 3.5.

3.1.3 Subroutine IVPT

This subroutine is used to generate the initial turbulent velocity profile for both models by specifying a Reynolds number based on momentum thickness, $R_\Theta \equiv u_e \Theta / \nu$ and local skin-friction coefficient c_f $[\equiv \tau_w / \frac{1}{2} \varrho u_e^2]$. It makes use of Eq. (1.2.9) for $y^+ \leq 50$; and Eq. (1.2.10) for $y^+ \geq 50$.

In terms of the Falkner-Skan variables, Eq. (1.2.9) can be written as

$$f' \sqrt{\frac{2}{c_f}} = \begin{cases} e_1 \eta & \eta \leq \dfrac{4}{e_1} \\[2mm] c_1 + c_2 \ln(e_1\eta) + c_3 [\ln(e_1\eta)]^2 + c_4 [\ln(e_1\eta)]^3 & \dfrac{4}{e_1} \leq \eta \leq \dfrac{50}{e_1} \end{cases} \tag{3.1.1}$$

where c_1, c_2, c_3 and c_4 are the coefficients of Eq. (1.2.9) and

$$e_1 = \sqrt{R_x} \sqrt{\frac{c_f}{2}}, \quad R_x = \frac{u_e x}{\nu} = R_L \bar{u}_e \xi, \quad R_L = \frac{u_\infty L}{\nu} \tag{3.1.2}$$

Similarly, for $\eta \geq 50/e_1$, we can use the expression given by Coles with Granville's correction for the whole profile [1], now with $\eta = y/\delta$,

$$u^+ = \frac{1}{\kappa} \ln y^+ + c + \frac{1}{\kappa} [\Pi(1 - \cos \pi\eta) + (\eta^2 - \eta^3)] \tag{3.1.3a}$$

Here Π is a profile parameter which is a constant equal to 0.55 for flows with zero pressure gradient provided that the momentum thickness Reynolds number R_Θ is greater than 5000. In terms of transformed variables, Eq. (3.1.3a) can be written as

$$f'\sqrt{\frac{2}{c_f}} = \frac{1}{\kappa}\ln(e_1\eta) + c + \frac{1}{\kappa}\left[\Pi\left(1 - \cos\pi\frac{\eta}{\eta_e}\right) + \left(\frac{\eta}{\eta_e}\right)^2 - \left(\frac{\eta}{\eta_e}\right)^3\right] \quad (3.1.3b)$$

where

$$\eta_e = \sqrt{R_x}\,\frac{\delta}{x} \quad (3.1.4)$$

It is clear that a complete velocity profile for a turbulent boundary-layer can be obtained from Eqs. (3.1.1) and (3.1.3) provided that the boundary-layer thickness δ and the profile parameter Π are known. Since they are not known at first, they must be calculated in a manner they are compatible with R_θ and c_f.

A convenient procedure is to assume δ^ν ($\nu = 0$) and, calculate Π from Eq. (3.1.3a) evaluated at the boundary-layer edge, $\eta = \eta_e$. The initial estimate of δ is obtained from the power-law relation,

$$\frac{\theta}{\delta} = \frac{n}{(1+n)(2+n)}$$

which for $n = 7$

$$\frac{\theta}{\delta} \sim 0.10$$

Here θ is calculated from the specified value of R_θ,

$$\theta = \frac{\nu}{u_e}R_\theta \quad (3.1.5)$$

The next values of δ^ν ($\nu = 1, 2, \ldots, n$) are obtained from

$$\delta^{(\nu+1)} = \delta^{(\nu)} - \frac{\phi}{\left(\frac{d\phi}{d\delta}\right)} \quad (3.1.6)$$

where, with ϕ_1 ($\equiv R_\theta/R_\delta$) given by [1]

$$\frac{R_\theta}{R_\delta} = \frac{u_\tau}{\kappa u_e}\left(\frac{11}{12} + \Pi\right) - \left(\frac{u_\tau}{\kappa u_e}\right)^2(1.9123016 + 3.0560\Pi + 1.5\Pi^2) \quad (3.1.7)$$

and θ denoting the momentum thickness calculated from Eq. (3.1.5),

$$\phi = \theta - \delta\phi_1 \quad (3.1.8a)$$

$$\frac{d\phi}{d\delta} = -\phi_1 - \delta\frac{d\phi_1}{d\Pi}\frac{d\Pi}{d\delta} \quad (3.1.8b)$$

and with

$$\frac{d\Pi}{d\delta} = -\frac{1}{2\delta} \quad (3.1.8c)$$

obtained by differentiating

$$\sqrt{\frac{2}{c_f}} \equiv \frac{u_e}{u_\tau} = \frac{1}{\kappa}\left[\ln\left(\frac{\delta u_e}{\nu}\frac{u_\tau}{u_e}\right) + 2\Pi\right] + c \quad (3.1.9)$$

with respect to δ.

3.1.4 Subroutine GROWTH

For most laminar-boundary-layer flows, the transformed boundary-layer thickness $\eta_e(x)$ is almost constant. A value of $\eta_e = 8$ is sufficient. However, for turbulent boundary-layers, $\eta_e(x)$ generally increases with increasing x. An estimate of $\eta_e(x)$ is determined by the following procedure.

We always require that $\eta_e(x^n) \geq \eta_e(x^{n-1})$, and in fact the calculations start with $\eta_e(x^0) = \eta_e(x_1)$. When the computations on $x = x^n$ (for any $n \geq 1$) have been completed, we test to see if $|v_J^n| \leq \varepsilon_v$ at $\eta_e(x^n)$ where, say $\varepsilon_v = 5 \times 10^{-4}$. This test is done in MAIN. If this test is satisfied, we set $\eta_e(x^{n+1}) = \eta_e(x^n)$. Otherwise, we call GROWTH and set $J_{\text{new}} = J_{\text{old}} + t$, where t is a number of points, say $t = 1$. In this case we also specify values of (f_j^n, u_j^n, v_j^n, b_j^n, k_j^n, ε_j^n etc.) for the new η_j points. We take the values of $u_j^n = 1$, $v_j^n = 0$, $f_j^n = (\eta_j - \eta_e)u_j^n + f_J^n$, $k_j^n = k_J^n$, $\varepsilon_j^n = \varepsilon_J^n$, $s_j^n = 0$, $q_j^n = 0$.

3.1.5 Subroutine GRID

The solution procedure for both models requires the generation of a grid normal to the surface, η-grid, and along the surface, x-grid. The latter requirement is satisfied by specifying locations with intervals which can be uniform or nonuniform. Its distribution depends on the variation of u_e with x so that the pressure gradient parameter $m(x)$ in the momentum equation can be calculated accurately. To ensure this requirement, it is necessary to take small Δx-steps (k_n) where there are rapid variations in $u_e(x)$ and where flow approaches separation.

For laminar flows, it is often sufficient to use a uniform grid in the η-direction. A choice of transformed boundary-layer thickness η_e equal to 8 often ensures that the dimensionless slope of the velocity profile at the edge, $f''(\eta_e)$, is sufficiently small ($< 10^{-3}$) and that approximately 41 j-points satisfies numerical accuracy requirements. For turbulent flows, however, a uniform grid is not satisfactory because the boundary-layer thickness η_e and dimensionless wall shear parameter f_w'' are much larger in turbulent flows than laminar flows. Since short steps in η must be taken to maintain computational accuracy when f_w'' is large, the steps near the wall in a turbulent boundary-layer must be shorter than the corresponding steps in a laminar boundary-layer under similar conditions.

A convenient and useful η-grid, discussed in [1] and used in this subroutine is a geometric progression having the property that the ratio of lengths of any two adjacent intervals is a constant; that is, $h_j = Kh_{j-1}$. The distance to the j-th line is given by the formula

$$\eta_j = h_1 \frac{K^j - 1}{K - 1}, \quad j = 1, 2, \ldots, J \quad K > 1 \tag{3.1.10}$$

There are two parameters: h_1, the length of the first $\Delta\eta$-step, and K, the ratio of two successive steps. The total number of points, J, can be calculated by the formula (see subroutine MYGRID):

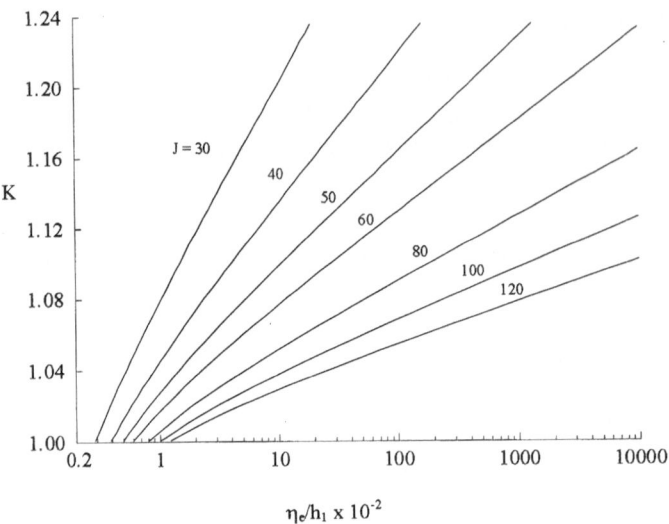

Fig. 3.1. Variation of K with h_1 for different η_e-values

$$J = \frac{\ln[1 + (K-1)(\eta_e/h_1)]}{\ln K} + 1 \qquad (3.1.11)$$

In most problems, calculations are performed by selecting h_1 and K and calculating the transformed boundary-layer thickness η_e. An idea about the number of points taken across the boundary-layer with the variable η-grid that uses those parameters for different η_e-values can be obtained from Fig. 3.1. For example, for $h_1 = 0.01$, $K = 1.10$ and $\eta_e = 100$, the ratio of η_e/h_1 is 10^4, and the number of points across the boundary-layer is approximately 70. Here in this program, we recommend K to be 1.12 (subroutine INPUT) which is sufficient for most turbulent flow calculations. Of course, if desired, this number can be changed with guidance from Fig. 3.1.

3.1.6 Subroutine OUTPUT

This subroutine prints out the desired profiles of the momentum, kinetic energy and rate of dissipation equations, such as f_j, u_j, v_j, k_j, ε_j as a function of η. It also computes the boundary-layer parameters, c_f, δ^*, θ, R_δ and R_θ.

3.2 Part 2 – CS Model

This part of the computer program which uses the CS model has five subroutines in addition to those described in Part 1. They include subroutines COEFTR, SOLV3, EDDY, GAMCAL, and CALFA and are briefly described in the following subsections.

3.2.1 Subroutine COEFTR

The solution of the momentum equation, Eq. (2.1.16), is much simpler than the solution of the k-ε model equations. Since this equation is third order, we have three first-order equations, the first two given by the first two equations in Eq. (2.2.1) and the third by Eq. (2.2.2). After writing the difference equations for Eqs. (2.2.1a,b) and linearizing them, we obtain

$$\delta f_j - \delta f_{j-1} - \frac{h_j}{2}(\delta u_j + \delta u_{j-1}) = (r_1)_j \tag{3.2.1a}$$

$$\delta u_j - \delta u_{j-1} - \frac{h_j}{2}(\delta v_j + \delta v_{j-1}) = (r_3)_{j-1} \tag{3.2.1b}$$

where

$$(r_1)_j = f_{j-1}^{(\nu)} - f_j^{(\nu)} + h_j u_{j-1/2}^{(\nu)} \tag{3.2.2a}$$

$$(r_3)_{j-1} = u_{j-1}^{(\nu)} - u_j^{(\nu)} + h_j v_{j-1/2}^{(\nu)} \tag{3.2.2b}$$

The third equation is given by

$$(s_1)_j \delta v_j + (s_2)_j \delta v_{j-1} + (s_3)_j \delta f_j + (s_4)_j \delta f_{j-1} + (s_5)_j \delta u_j + (s_6)_j \delta u_{j-1} = (r_2)_j \tag{3.2.2c}$$

with $(r_2)_j$ given by Eq. (2.3.14).

The linearized boundary conditions correspond to Eqs. (2.3.4a,b) at $\eta = 0$ and to $\delta u_J = 0$ at $\eta = \eta_J$. This system of equations, as described in subsection 2.3.1, is again written in matrix-vector form given by Eq. (2.3.15) with A_j, B_j and C_j matrices given by

$$A_0 = \begin{vmatrix} 1 & 0 & 0 \\ 0 & 1 & 0 \\ 0 & -1 & -h_1/2 \end{vmatrix} \qquad A_j \equiv \begin{vmatrix} 1 & -h_j/2 & 0 \\ (s_3)_j & (s_5)_j & (s_1)_j \\ 0 & -1 & -h_{j+1}/2 \end{vmatrix} \qquad 1 \le j \le J-1 \tag{3.2.3a}$$

$$A_J \equiv \begin{vmatrix} 1 & -h_J/2 & 0 \\ (s_3)_J & (s_5)_J & (s_1)_J \\ 0 & 1 & 0 \end{vmatrix} \qquad B_j = \begin{vmatrix} -1 & -h_j/2 & 0 \\ (s_4)_j & (s_6)_j & (s_2)_j \\ 0 & 0 & 0 \end{vmatrix} \qquad 1 \le j \le J \tag{3.2.3b}$$

$$C_j \equiv \begin{vmatrix} 0 & 0 & 0 \\ 0 & 0 & 0 \\ 0 & 1 & -h_{j+1}/2 \end{vmatrix} \qquad 0 \le j \le J-1 \tag{3.2.3c}$$

and $\vec{\delta}_j$ and $\vec{r}_j$ by

$$\vec{\delta}_j = \begin{vmatrix} \delta f_j \\ \delta u_j \\ \delta v_j \end{vmatrix} \qquad \vec{r}_j = \begin{vmatrix} (r_1)_j \\ (r_2)_j \\ (r_2)_j \end{vmatrix} \qquad 0 \le j \le J \tag{3.2.4}$$

The solution of the Eq. (2.3.15) is again obtained with the block elimination method described in subsection 2.3.4.

This subroutine contains the coefficients of the linearized momentum equation given by Eqs. (3.2.2c), (3.2.1) and (2.3.14). Since the calculations are for turbulent flow only, these coefficients for the first two computed x-stations are slightly different due to the use of two-point backward difference formulas (see subsection 2.3.1) for the streamwise derivatives in the momentum equation. This is needed to avoid oscillations caused by the specified initial velocity profiles. At the third x-station, the calculations revert back to the central differences for the streamwise derivatives described in [1]. In this case the coefficients $(s_1)_j$ to $(s_6)_j$ and $(r_2)_j$ are given by

$$(s_1)_j = h_j^{-1} b_j^{(\nu)} + \frac{\alpha_1}{2} f_j^{(\nu)} + \frac{\alpha^n}{2} f_{j-1/2}^{(\nu)} \tag{3.2.5a}$$

$$(s_2)_j = -h_j^{-1} b_{j-1}^{(\nu)} + \frac{\alpha_1}{2} f_{j-1}^{(\nu)} + \frac{\alpha^n}{2} f_{j-1/2}^{(\nu)} \tag{3.2.5b}$$

$$(s_3)_j = \frac{\alpha_1}{2} v_j^{(\nu)} + \frac{\alpha^n}{2} v_{j-1/2}^{(\nu)} \tag{3.2.5c}$$

$$(s_3)_j = \frac{\alpha_1}{2} v_{j-1}^{(\nu)} + \frac{\alpha^n}{2} v_{j-1/2}^{(\nu)} \tag{3.2.5d}$$

$$(s_5)_j = -\alpha_2 u_j^{(\nu)} \tag{3.2.5e}$$

$$(s_6)_j = -\alpha_2 u_{j-1}^{(\nu)} \tag{3.2.5f}$$

$$(r_2)_j = R_{j-1/2}^{n-1} - \left[\begin{array}{l} h_j^{-1}(b_j^{(\nu)} v_j^{(\nu)} - b_{j-1}^{(\nu)} v_{j-1}^{(\nu)}) + \alpha_1(fv)_{j-1/2}^{(\nu)} \\ - \alpha_2(u^2)_{j-1/2}^{(\nu)} - \alpha^n(v_{j-1/2}^{n-1} f_{j-1/2}^{(\nu)} - f_{j-1/2}^{n-1} v_{j-1/2}^{(\nu)}) \end{array} \right] \tag{3.2.6}$$

where

$$R_{j-1/2}^{n-1} = -L_{j-1/2}^{n-1} + \alpha^n [(fv)_{j-1/2}^{n-1} - (u^2)_{j-1/2}^{n-1}] - m^n \tag{3.2.7a}$$

$$L_{j-1/2}^{n-1} = \left\{ h_j^{-1}(b_j v_j - b_{j-1} v_{j-1}) + m_1(fv)_{j-1/2} + m[1 - (u^2)_{j-1/2}] \right\}^{n-1} \tag{3.2.7b}$$

$$\alpha_1 = m_1 + \alpha, \qquad \alpha_2 = m + \alpha$$

3.2.2 Subroutine SOLV3

This subroutine is used to solve the linear system given by Eq. (2.3.15) with the block elimination method discussed in subsection 2.3.4. To describe the procedure for obtaining the recursion formulas used in this subroutine we first consider Eq. (2.3.30). Noting that the Γ_j matrix has the same structure as B_j and denoting the elements of γ_j, by γ_{ik} $(i, k = 1, 2, 3)$, we can write Γ_j as

$$\Gamma_j \equiv \begin{vmatrix} (\gamma_{11})_j & (\gamma_{12})_j & (\gamma_{13})_j \\ (\gamma_{21})_j & (\gamma_{22})_j & (\gamma_{23})_j \\ 0 & 0 & 0 \end{vmatrix} \tag{3.2.8a}$$

Similarly, if the elements of Δ_j are denoted by α_{ik} we can write Δ_j as [note that the third row of Δ_j follows from the third row of A_j according to Eq. (2.3.30c)]

$$\Delta_j \equiv \begin{vmatrix} (\alpha_{11})_j & (\alpha_{12})_j & (\alpha_{13})_j \\ (\alpha_{21})_j & (\alpha_{22})_j & (\alpha_{23})_j \\ 0 & -1 & -h_{j+1}/2 \end{vmatrix} \quad 0 \le j \le J-1 \tag{3.2.8b}$$

and for $j = J$, the first two rows are the same as the first two rows in Eq. (2.3.30b), but the elements of the third row, which correspond to the boundary conditions at $j = J$, are $(0, 1, 0)$.

For $j = 0$, $\Delta_0 = A_0$; therefore the values of $(\alpha_{ik})_0$ are

$$\begin{aligned} (\alpha_{11})_0 &= 1 & (\alpha_{12})_0 &= 0 & (\alpha_{13})_0 &= 0 \\ (\alpha_{21})_0 &= 0 & (\alpha_{22})_0 &= 1 & (\alpha_{23})_0 &= 0 \end{aligned} \tag{3.2.9a}$$

and the values of $(\gamma_{ik})_1$ are

$$\begin{aligned} (\gamma_{11})_1 &= -1 & (\gamma_{12})_1 &= -\frac{1}{2}h_1 & (\gamma_{13})_1 &= 0 \\ (\gamma_{21})_1 &= (s_4)_1 & (\gamma_{23})_1 &= -2\left[\frac{(s_2)_1}{h_1}\right] & (\gamma_{22})_1 &= (s_6)_1 + (\gamma_{23})_1 \end{aligned} \tag{3.2.9b}$$

The elements of the Δ_j matrices are calculated from Eq. (2.3.30a). Using the definitions of A_j, Γ_j and C_{j-1}, we find from Eq. (2.3.30c) that for $j = 1, 2, \ldots, J$,

$$\begin{aligned} (\alpha_{11})_j &= 1 & (\alpha_{12})_j &= -\frac{h_j}{2} - (\gamma_{13})_j & (\alpha_{13})_j &= \frac{h_j}{2}(\gamma_{13})_j \\ (\alpha_{21})_j &= (s_3)_j & (\alpha_{22})_j &= (s_5)_j - (\gamma_{23})_j & (\alpha_{23})_j &= (s_1)_j + \frac{h_j}{2}(\gamma_{23})_j \end{aligned}$$
$$\tag{3.2.10a}$$

To find the elements of the Γ_j matrices, we use Eq. (2.3.30b). With Δ_j defined by Eq. (3.2.8b) and B_j by Eq. (3.2.3b), it follows that for $1 \le j \le J$,

$$(\gamma_{11})_j = \left\{ (\alpha_{23})_{j-1} + \frac{h_j}{2} \left[\left(\frac{h_j}{2} \right) (\alpha_{21})_{j-1} - (\alpha_{22})_{j-1} \right] \right\} \Big/ \Delta_0$$

$$(\gamma_{12})_j = -\left\{ \frac{h_j}{2}\frac{h_j}{2} + (\gamma_{11})_j \left[(\alpha_{12})_{j-1}\frac{h_j}{2} - (\alpha_{13})_{j-1} \right] \right\} \Big/ \Delta_0$$

$$(\gamma_{13})_j = [(\gamma_{11})_j(\alpha_{13})_{j-1} + (\gamma_{12})(\alpha_{23})_{j-1}]/\frac{h_j}{2}$$

$$(\gamma_{21})_j = \left\{ (s_2)_j(\alpha_{21})_{j-1} - (s_4)_j(\alpha_{23})_{j-1} \right.$$
$$\left. + \frac{h_j}{2}[(s_4)_j(\alpha_{22})_{j-1} - (s_6)_j(\alpha_{21})_{j-1}] \right\} \Big/ \Delta_0$$

$$(\gamma_{22})_j = \left\{ (s_6)_j\frac{h_j}{2} - (s_2)_j + (\gamma_{21})_j \left[(\alpha_{13})_{j-1} - (\alpha_{12})_{j-1}\frac{h_j}{2} \right] \right\} \Big/ \Delta_1 \quad \text{(3.2.10b)}$$

$$(\gamma_{23})_j = (\gamma_{21})_j(\alpha_{12})_{j-1} + (\gamma_{22})_j(\alpha_{22})_{j-1} - (s_6)_j$$

$$\Delta_0 = (\alpha_{13})_{j-1}(\alpha_{21})_{j-1} - (\alpha_{23})_{j-1}(\alpha_{11})_{j-1}$$
$$- \frac{h_j}{2}[(\alpha_{12})_{j-1}(\alpha_{21})_{j-1} - (\alpha_{22})_{j-1}(\alpha_{11})_{j-1}]$$

$$\Delta_1 = (\alpha_{22})_{j-1}\frac{h_j}{2} - (\alpha_{23})_{j-1}$$

To summarize the calculation of Γ_j and Δ_j matrices, we first calculate α_{ik} from Eq. (3.2.9a) for $j = 0$, γ_{ik} from Eq. (3.2.9b) for $j = 1$, α_{ik} from Eq. (3.2.10a) for $j = 1$, then γ_{ik} from Eq. (3.2.10b) for $j = 2$, α_{ik} from Eq. (3.2.10a) for $j = 2$, then γ_{ik} from Eq. (3.2.10b) for $j = 3$, etc.

In the second part of the forward sweep we compute $\vec{w}_j$ from the relations given by Eq. (2.3.31). If we denote the components of the vector $\vec{w}_j$ by

$$\vec{w}_j \equiv \begin{vmatrix} (w_1)_j \\ (w_2)_j \\ (w_3)_j \end{vmatrix} \quad 0 \le j \le J \qquad \text{(3.2.11)}$$

Then it follows from Eq. (2.3.31a) that for $j = 0$,

$$(w_1)_0 = (r_1)_0 \quad (w_2)_0 = (r_2)_0 \quad (w_3)_0 = (r_3)_0 \qquad \text{(3.2.12a)}$$

and from Eq. (2.3.31b) for $1 \le j \le J$,

$$(w_1)_j = (r_1)_j - (\gamma_{11})_j(w_1)_{j-1} - (\gamma_{12})_j(w_2)_{j-1} - (\gamma_{13})_j(w_3)_{j-1}$$
$$(w_2)_j = (r_2)_j - (\gamma_{21})_j(w_1)_{j-1} - (\gamma_{22})_j(w_2)_{j-1} - (\gamma_{23})_j(w_3)_{j-1} \qquad \text{(3.2.12b)}$$
$$(w_3)_j = (r_3)_j$$

In the backward sweep, $\vec{\delta}_j$ is computed from the formulas given by Eq. (2.3.32). With the definitions of $\vec{\delta}_j$, Δ_j and $\vec{w}_j$, it follows from Eq. (2.3.32a) that

$$\delta u_J = (w_3)_J \qquad \text{(3.2.13a)}$$

$$\delta v_J = \frac{e_2(\alpha_{11})_J - e_1(\alpha_{21})_J}{(\alpha_{23})_J(\alpha_{11})_J - (\alpha_{13})_J(\alpha_{21})_J} \tag{3.2.13b}$$

$$\delta f_J = \frac{e_1 - (\alpha_{13})_J \delta v_J}{(\alpha_{11})_J} \tag{3.2.13c}$$

where

$$e_1 = (w_1)_J - (\alpha_{12})_J \delta u_J$$
$$e_2 = (w_2)_J - (\alpha_{22})_J \delta u_J$$

The components of $\vec{\delta}$, for $j = J-1, J-2, \ldots, 0$, follow from Eq. (2.3.32b)

$$\delta v_j = \frac{(\alpha_{11})_j[(w_2)_j + e_3(\alpha_{22})_j] - (\alpha_{21})_j(w_1)_j - e_3(\alpha_{21})_j(\alpha_{12})_j}{\Delta_2} \tag{3.2.14a}$$

$$\delta u_j = -\frac{h_{j+1}}{2}\delta v_j - e_3 \tag{3.2.14b}$$

$$\delta f_j = \frac{(w_1)_j - (\alpha_{12})_j \delta u_j - (\alpha_{13})_j \delta v_j}{(\alpha_{11})_j} \tag{3.2.14c}$$

where

$$e_3 = (w_3)_j - \delta u_{j+1} + \frac{h_{j+1}}{2}\delta v_{j+1}$$
$$\Delta_2 = (\alpha_{21})_j(\alpha_{12})_j\frac{h_{j+1}}{2} - (\alpha_{21})_j(\alpha_{13})_j \tag{3.2.14d}$$
$$- \frac{h_{j+1}}{2}(\alpha_{22})_j(\alpha_{11})_j + (\alpha_{23})_j(\alpha_{11})_j$$

To summarize, one iteration of Newton's method is carried out as follows. The vectors $\vec{r}_j$ defined in Eq. (2.3.15) are computed from Eqs. (2.3.14) and (3.2.2) by using the latest iterate. The matrix elements of A_j, B_j and C_j defined in Eq. (3.2.3) are next determined by Eq. (3.2.5a) to (3.2.5f). Using the relations in Eqs. (2.3.30) and (2.3.31), the matrices Γ_j and Δ_j and vectors $\vec{w}_j$ are calculated. The matrix elements for Γ_j defined in Eq. (2.3.30b) are determined from Eqs. (3.2.9b) and (3.2.10b). The components of the vector $\vec{w}_j$ defined in Eq. (3.2.11) are determined from Eq. (3.2.12). In the backward sweep, the components of $\vec{\delta}_j$ are computed from Eqs. (3.2.13) and (3.2.14).

3.2.3 Subroutines EDDY, GAMCAL, CALFA

These subroutines use the CS algebraic eddy viscosity formulation discussed in Section 1.1. In terms of transformed variables, $(\varepsilon_m^+)_i$ and $(\varepsilon_m^+)_0$ are given by

$$(\varepsilon_m^+)_i = 0.16\eta^2\sqrt{R_x}v\left\{1 - \exp(-R_x^{1/4}v_w^{1/2}/26/c_n)\right\}\gamma_{tr} \tag{3.2.15a}$$

$$(\varepsilon_m^+)_0 = \alpha\sqrt{R_x}(\eta_J - f_J)\gamma_{tr}\gamma \tag{3.2.15b}$$

where

$$c_n = \frac{m}{R_x^{1/4} v_w^{3/4}} \qquad (3.2.16)$$

Subroutine EDDY contains the expressions for the inner and outer regions. The intermittency expression used in the outer eddy viscosity formula, Eq. (1.1.8) is calculated in subroutine GAMCAL and the variable α given by Eq. (1.1.9) in subroutine CALFA.

3.3 Part 3 – k-ε Model

The structure of the k-ε model, which includes the zonal method and the model for low and high Reynolds number flows, is similar to the CS model described in the previous section. It consists of the subroutines described below.

3.3.1 Subroutines KECOEF, KEPARM, KEDEF and KEDAMP

Again we need a subroutine for the coefficients of the linearized equations for momentum, turbulent kinetic energy and rate of dissipation. We also need to generate initial profiles for the kinetic energy and rate of dissipation equations for both low and high Reynolds number flows. We do not need to generate the initial turbulent velocity profile for the momentum equation since it is already generated by subroutine IVPT discussed in Part 1. Then we need an algorithm, like SOLV3, to solve the linear system of equations for the zonal method and k-ε model with and without wall functions for low and high Reynolds number flows.

To simplify the coding and discussion and the application of this computer program to other turbulence models, we use three additional subroutines to define the coefficients of the linearized equations for momentum, kinetic energy and rate of dissipation given in subroutine KECOEF. The first of these three subroutines is subroutine KEPARM, which calculates the parameters b_1, b_2, b_3 and production and dissipation terms and their linearized terms such as $\left(\frac{\partial b_2}{\partial k}\right)_j^n$, $\left(\frac{\partial b_3}{\partial \varepsilon}\right)_j^n$, $\left(\frac{\partial P}{\partial \varepsilon}\right)_j^n$, $\left(\frac{\partial Q}{\partial k}\right)_j^n$, $\left(\frac{\partial P}{\partial v}\right)_j^n$, etc. in the equations for kinetic energy and rate of dissipation.

The second of these three subroutines is subroutine KEDEF, which calculates D, E, F terms, (see Section 1.2), and their linearized terms such as $\left(\frac{\partial E}{\partial \varepsilon}\right)_j^n$, $\left(\frac{\partial F}{\partial \varepsilon}\right)_j^n$, etc. in the k-ε model associated with low Reynolds number effects, which in the present program correspond to the models of Huang-Lin and Chien discussed in Section 1.2.

The third of these three subroutines is subroutine KEDAMP, which calculates near-wall damping terms f_1, f_2, f_μ, σ_k, σ_ε and their linearized terms which are for low Reynolds numbers and are model dependent.

The linearized coefficients of the momentum equation in subroutine KECOEF use both two and three point backward finite-difference approximations for the streamwise derivatives. For $j \leq j_s$, the coefficients $(s_1)_j$ to $(s_6)_j$ are given by Eq. (3.2.5) for the CS model. At $j = j_s$

$$(\varepsilon_m)_{CS} = (\varepsilon_m)_{k\text{-}\varepsilon}$$

and $(s_7)_j$ to $(s_{12})_j$ are given by the following equations,

$$(s_7)_j = h_j^{-1} v_j \left(\frac{\partial b}{\partial k} \right)_j \tag{3.3.1a}$$

$$(s_8)_j = -h_j^{-1} v_{j-1} \left(\frac{\partial b}{\partial k} \right)_{j-1} \tag{3.3.1b}$$

$$(s_9)_j = h_j^{-1} v_j \left(\frac{\partial b}{\partial s} \right)_j \tag{3.3.1c}$$

$$(s_{10})_j = -h_j^{-1} v_{j-1} \left(\frac{\partial b}{\partial s} \right)_{j-1} \tag{3.3.1d}$$

$$(s_{11})_j = h_j^{-1} v_j \left(\frac{\partial b}{\partial \varepsilon} \right)_j \tag{3.3.1e}$$

$$(s_{12})_j = -h_j^{-1} v_{j-1} \left(\frac{\partial b}{\partial \varepsilon} \right)_{j-1} \tag{3.3.1f}$$

for the k-ε model.

This subroutine also presents the coefficients of the kinetic energy equation, $(\alpha_1)_j$ to $(\alpha_{12})_j$ and $(r_4)_j$ in Eq. (2.3.25) and the coefficients of the rate of dissipation equation, $(\beta_1)_j$ to $(\beta_{14})_j$ and $(\tau_5)_j$ in Eq. (2.3.26).

To discuss the procedure for obtaining the coefficients of the kinetic energy and rate of dissipation equations, consider Eq. (2.2.3). With x-wise derivatives represented either by two- or three-point backward differences, the finite difference approximations to Eq. (2.2.3) are

$$h_j^{-1}[(b_2 s)_j^n - (b_2 s)_{j-1}^n] + m_1^n (f s)_{j-1/2}^n - 2m^n (uk)_{j-1/2}^n + P_{j-1/2}^n$$

$$-Q_{j-1/2}^n + F_{j-1/2}^n = x^n \left[u_{j-1/2}^n \left(\frac{\partial k}{\partial x} \right)_{j-1/2}^n - s_{j-1/2}^n \left(\frac{\partial f}{\partial x} \right)_{j-1/2}^n \right] \tag{3.3.2}$$

Linearizing, we get

$$h_j^{-1} \left[(b_2)_j^n \delta s_j + s_j^n \left(\frac{\partial b_2}{\partial k} \right)_j^n \delta k_j + s_j^n \left(\frac{\partial b_2}{\partial \varepsilon} \right)_j^n \delta \varepsilon_j - (b_2)_{j-1}^n \delta s_{j-1} \right.$$

$$\left. - s_{j-1}^n \left(\frac{\partial b_2}{\partial k} \right)_{j-1}^n \delta k_{j-1} - s_{j-1}^n \left(\frac{\partial b_2}{\partial \varepsilon} \right)_{j-1}^n \delta \varepsilon_{j-1} \right]$$

$$+ \frac{1}{2}\left[\left(\frac{\partial P}{\partial k}\right)^n_j \delta k_j + \left(\frac{\partial P}{\partial \varepsilon}\right)^n_j \delta\varepsilon_j + \left(\frac{\partial P}{\partial v}\right)^n_j \delta v_j \right.$$

$$\left. + \left(\frac{\partial P}{\partial k}\right)^n_{j-1} \delta k_{j-1} + \left(\frac{\partial P}{\partial \varepsilon}\right)^n_{j-1} \delta\varepsilon_{j-1} + \left(\frac{\partial P}{\partial v}\right)^n_{j-1} \delta v_{j-1}\right]$$

$$+ \frac{m_1^n}{2}(f_j^n \delta s_j + s_j^n \delta f_j + f_{j-1}^n \delta s_{j-1} + s_{j-1}^n \delta f_{j-1})$$

$$- m^n[u_j^n \delta k_j + k_j^n \delta u_j + u_{j-1}^n \delta k_{j-1} + k_{j-1}^n \delta u_{j-1}]$$

$$- \frac{1}{2}\left[\left(\frac{\partial Q}{\partial k}\right)_j \delta k_j + \left(\frac{\partial Q}{\partial \varepsilon}\right)_j \delta\varepsilon_j + \left(\frac{\partial Q}{\partial k}\right)_{j-1} \delta k_{j-1} + \left(\frac{\partial Q}{\partial \varepsilon}\right)_{j-1} \delta\varepsilon_{j-1}\right]$$

$$+ \frac{1}{2}\left[\left(\frac{\partial F}{\partial k}\right)_j \delta k_j + \left(\frac{\partial F}{\partial \varepsilon}\right)_j \delta\varepsilon_j + \left(\frac{\partial F}{\partial k}\right)_{j-1} \delta k_{j-1} + \left(\frac{\partial F}{\partial \varepsilon}\right)_{j-1} \delta\varepsilon_{j-1}\right]$$

$$= \frac{x^n}{2}\left\{(\delta u_j + \delta u_{j-1})\left(\frac{\partial k}{\partial x}\right)^n_{j-1/2} + u_{j-1/2}^n\left[\frac{\partial}{\partial k}\left(\frac{\partial k}{\partial x}\right)^n_j \delta k_j\right.\right.$$

$$\left. + \frac{\partial}{\partial k}\left(\frac{\partial k}{\partial x}\right)^n_{j-1} \delta k_{j-1}\right] - (\delta s_j + \delta s_{j-1})\left(\frac{\partial f}{\partial x}\right)^n_{j-1/2}$$

$$\left. - s_{j-1/2}^n\left[\frac{\partial}{\partial f}\left(\frac{\partial f}{\partial x}\right)^n_j \delta f_j + \frac{\partial}{\partial f}\left(\frac{\partial f}{\partial x}\right)^n_{j-1} \delta f_{j-1}\right]\right\} + (r_4)_j \quad (3.3.3)$$

The coefficients of Eq. (2.3.25) can be written as

$$(\alpha_1)_j = \frac{m_1^n}{2}s_j^n + \frac{x^n}{2}s_{j-1/2}^n\frac{\partial}{\partial f}\left(\frac{\partial f}{\partial x}\right)^n_j \tag{3.3.4a}$$

$$(\alpha_2)_j = \frac{m_1^n}{2}s_{j-1}^n + \frac{x^n}{2}s_{j-1/2}^n\frac{\partial}{\partial f}\left(\frac{\partial f}{\partial x}\right)^n_{j-1} \tag{3.3.4b}$$

$$(\alpha_3)_j = -m^n k_j^n - \frac{x^n}{2}\left(\frac{\partial k}{\partial x}\right)^n_{j-1/2} \tag{3.3.4c}$$

$$(\alpha_4)_j = -m^n k_{j-1}^n - \frac{x^n}{2}\left(\frac{\partial k}{\partial x}\right)_{j-1/2} \tag{3.3.4d}$$

$$(\alpha_5)_j = \frac{1}{2}\left(\frac{\partial P}{\partial v}\right)^n_j \tag{3.3.4e}$$

$$(\alpha_6)_j = \frac{1}{2}\left(\frac{\partial P}{\partial v}\right)^n_{j-1} \tag{3.3.4f}$$

$$(\alpha_7)_j = h_j^{-1}s_j^n\left(\frac{\partial b_2}{\partial k}\right)^n_j + \frac{1}{2}\left(\frac{\partial P}{\partial k}\right)_j - m^n u_j^n - \frac{1}{2}\left(\frac{\partial Q}{\partial k}\right)^n_j$$

$$+ \frac{1}{2}\left(\frac{\partial F}{\partial k}\right)^n_j - \frac{x^n}{2}u_{j-1/2}^n\frac{\partial}{\partial k}\left(\frac{\partial k}{\partial x}\right)^n_j \tag{3.3.4g}$$

$$(\alpha_8)_j = -h_j^{-1} s_{j-1}^n \left(\frac{\partial b_2}{\partial k}\right)_{j-1}^n + \frac{1}{2}\left(\frac{\partial P}{\partial k}\right)_{j-1} - m^n u_{j-1}^n$$

$$- \frac{1}{2}\left(\frac{\partial Q}{\partial k}\right)_{j-1}^n + \frac{1}{2}\left(\frac{\partial F}{\partial k}\right)_{j-1}^n - \frac{x^n}{2} u_{j-1/2}^n \frac{\partial}{\partial k}\left(\frac{\partial k}{\partial x}\right)_{j-1}^n \tag{3.3.4h}$$

$$(\alpha_9)_j = h_j^{-1}(b_2)_j^n + \frac{m_1^n}{2} f_j + \frac{x^n}{2}\left(\frac{\partial f}{\partial x}\right)_{j-1/2}^n \tag{3.3.4i}$$

$$(\alpha_{10})_j = -h_j^{-1}(b_2)_{j-1}^n + \frac{m_1^n}{2} f_{j-1} + \frac{x^n}{2}\left(\frac{\partial f}{\partial x}\right)_{j-1/2}^n \tag{3.3.4j}$$

$$(\alpha_{11})_j = h_j^{-1} s_j^n \left(\frac{\partial b_2}{\partial \varepsilon}\right)_j^n + \frac{1}{2}\left[\left(\frac{\partial P}{\partial \varepsilon}\right)_j^n - \left(\frac{\partial Q}{\partial \varepsilon}\right)_j^n + \left(\frac{\partial F}{\partial \varepsilon}\right)_j^n\right] \tag{3.3.4k}$$

$$(\alpha_{12})_j = -h_j^{-1} s_{j-1}^n \left(\frac{\partial b_2}{\partial \varepsilon}\right)_{j-1}^n + \frac{1}{2}\left[\left(\frac{\partial P}{\partial \varepsilon}\right)_{j-1}^n - \left(\frac{\partial Q}{\partial \varepsilon}\right)_{j-1}^n + \left(\frac{\partial F}{\partial \varepsilon}\right)_{j-1}^n\right] \tag{3.3.4l}$$

$$(r_4)_j = -[(b_2 s)_j^n - (b_2 s)_{j-1}^n] h_j^{-1}$$
$$- [P_{j-1/2}^n - Q_{j-1/2}^n + F_{j-1/2}^n - 2m^n(uk)_{j-1/2}^n$$
$$+ m_1^n(fs)_{j-1/2}^n] \tag{3.3.5a}$$
$$+ x^n \left[u_{j-1/2}^n \left(\frac{\partial k}{\partial x}\right)_{j-1/2}^n - s_{j-1/2}^n \left(\frac{\partial f}{\partial x}\right)_{j-1/2}^n\right]$$

Remembering the definitions of the diffusion, production, dissipation, convection and F terms, Eq. (2.9.4a) can also be written as

$$(r_4)_j = -[\text{diffusion} + \text{production} - \text{dissipation} + F - \text{convection}] \tag{3.3.5b}$$

The parameter P, Q and F are model-dependent. As a result, the derivatives with respect to k, ε and v will be different for each model. The derivations of $\frac{\partial k}{\partial x}$ and $\frac{\partial f}{\partial x}$ with respect to k and f are straightforward.

In term of transformed variables the parameters P, Q and F in Huang and Lin's model, for example, are (here ε is $\bar{\varepsilon}$)

$$P = c_\mu f_\mu R_x \frac{k^2}{\varepsilon} v^2 \tag{3.3.6a}$$

$$Q = \varepsilon + \frac{1}{2}\frac{s^2}{k}, \qquad D = \frac{1}{2}\frac{s^2}{k} \tag{3.3.6b}$$

$$F = \left[\frac{1}{2Q}\left(ss' - \frac{1}{2}\frac{s^3}{k}\right)\right]' \tag{3.3.6c}$$

We now consider the rate of dissipation equation given by Eq. (2.2.4). Following a procedure similar to the one used for the kinetic energy equation, the finite-difference approximation for Eq. (2.2.4) is

$$h_j^{-1}[(b_3q)_j^n - (b_3q)_{j-1}^n] + (P_1)_{j-1/2}^n - (Q_1)_{j-1/2}^n + (E)_{j-1/2}^n$$
$$+ m_1^n(fq)_{j-1/2}^n - (3m^n - 1)(u\varepsilon)_{j-1/2}^n \tag{3.3.7}$$
$$= x\left[u_{j-1/2}^n\left(\frac{\partial\varepsilon}{\partial x}\right)_{j-1/2}^n - q_{j-1/2}^n\left(\frac{\partial f}{\partial x}\right)_{j-1/2}^n\right]$$

After linearization, the resulting expression can be expressed in the form given by Eq. (2.3.26),

$$(\beta_1)_j = \frac{m_1^n}{2}q_j^n + \frac{x^n}{2}q_{j-1/2}^n\frac{\partial}{\partial f}\left(\frac{\partial f}{\partial x}\right)_j^n \tag{3.3.8a}$$

$$(\beta_2)_j = \frac{m_1^n}{2}q_{j-1}^n + \frac{x^n}{2}q_{j-1/2}^n\frac{\partial}{\partial f}\left(\frac{\partial f}{\partial x}\right)_{j-1}^n \tag{3.3.8b}$$

$$(\beta_3)_j = -\frac{x^n}{2}\left(\frac{\partial\varepsilon}{\partial x}\right)_{j-1/2}^n - \frac{1}{2}(3m - 1)\varepsilon_j^n \tag{3.3.8c}$$

$$(\beta_4)_j = -\frac{x^n}{2}\left(\frac{\partial\varepsilon}{\partial x}\right)_{j-1/2}^n - \frac{1}{2}(3m - 1)\varepsilon_{j-1}^n \tag{3.3.8d}$$

$$(\beta_5)_j = \frac{1}{2}\left(\frac{\partial P_1}{\partial v}\right)_j^n \tag{3.3.8e}$$

$$(\beta_6)_j = \frac{1}{2}\left(\frac{\partial P_1}{\partial v}\right)_{j-1/2}^n \tag{3.3.8f}$$

$$(\beta_7)_j = h_j^{-1}q_j^n\left(\frac{\partial b_3}{\partial k}\right)_j^n + \frac{1}{2}\left(\frac{\partial P_1}{\partial k}\right)_j^n - \frac{1}{2}\left(\frac{\partial Q_1}{\partial k}\right)_j^n \tag{3.3.8g}$$

$$(\beta_8)_j = -h_j^{-1}q_{j-1}^n\left(\frac{\partial b_3}{\partial k}\right)_{j-1}^n + \frac{1}{2}\left(\frac{\partial P_1}{\partial k}\right)_{j-1}^n - \frac{1}{2}\left(\frac{\partial Q_1}{\partial k}\right)_{j-1}^n \tag{3.3.8h}$$

$$(\beta_{11})_j = h_j^{-1}q_j^n\left(\frac{\partial b_3}{\partial \varepsilon}\right)_j^n - \frac{1}{2}\left(\frac{\partial Q_1}{\partial \varepsilon}\right)_j^n + \frac{1}{2}\left(\frac{\partial E}{\partial \varepsilon}\right)_j^n$$
$$- \frac{1}{2}(3m^n - 1)u_j^n - \frac{x^n}{2}u_{j-1/2}^n\frac{\partial}{\partial \varepsilon}\left(\frac{\partial\varepsilon}{\partial x}\right)_j^n \tag{3.3.8i}$$

$$(\beta_{12})_j = -h_j^{-1}q_{j-1}^n\left(\frac{\partial b_3}{\partial \varepsilon}\right)_{j-1}^n - \frac{1}{2}\left(\frac{\partial Q_1}{\partial \varepsilon}\right)_{j-1}^n + \frac{1}{2}\left(\frac{\partial E}{\partial \varepsilon}\right)_{j-1}^n$$
$$- \frac{1}{2}(3m^n - 1)u_{j-1}^n - \frac{x^n}{2}u_{j-1/2}^n\frac{\partial}{\partial \varepsilon}\left(\frac{\partial\varepsilon}{\partial x}\right)_{j-1}^n \tag{3.3.8j}$$

$$(\beta_{13})_j = h_j^{-1}(b_3)_j^n + \frac{m_1^n}{2}f_j^n + \frac{x^n}{2}\left(\frac{\partial f}{\partial x}\right)_{j-1/2}^n \tag{3.3.8k}$$

$$(\beta_{14})_j = -h_j^{-1}(b_3)_{j-1}^n + \frac{m_1^n}{2}f_{j-1}^n + \frac{x^n}{2}\left(\frac{\partial f}{\partial x}\right)_{j-1/2}^n \tag{3.3.8l}$$

$$(r_5)_j = -h_j^{-1}[(b_3q)_j^n - (b_3q)_{j-1}^n]$$

$$- [(P_1)_{j-1/2}^n - (Q_1)_{j-1/2}^n + E_{j-1/2}^n$$

$$+ m_1^n(fq)_{j-1/2}^n - (3m^n - 1)(u\varepsilon)_{j-1/2}^n] \tag{3.3.9a}$$

$$+ x^n \left[u_{j-1/2}^n \left(\frac{\partial \varepsilon}{\partial x} \right)_{j-1/2}^n - q_{j-1/2}^n \left(\frac{\partial f}{\partial x} \right)_{j-1/2}^n \right]$$

Equation (3.3.9a) can also be written are

$$(r_5)_j = -[\text{diffusion} + \text{generation} - \text{destruction} - \text{convection} + E] \tag{3.3.9b}$$

3.3.2 Subroutine KEINITK

This subroutine generates the initial k-profile for low and high Reynolds numbers as well as the profile for the zonal method. For high Reynolds number flows, the kinetic energy profile k is determined by first calculating the shear stress τ from

$$\tau = (\varepsilon_m)_{CS} \frac{\partial u}{\partial y} \tag{3.3.10}$$

and using the relation between τ and k,

$$k = \frac{\tau}{a_1} \tag{3.3.11}$$

with $a_1 = 0.30$. The calculation of τ is easily accomplished in subroutine IVPT once the initial velocity profile is generated in that subroutine.

For low Reynolds number flows, we assume that the ratio of τ^+/k^+ is given by

$$\frac{\tau^+}{k^+} = \begin{cases} a(y^+)^2 + b(y^+)^3 & y^+ \le 4.0 & (3.3.12a) \\ c_1 + c_2z + c_3z^2 + c_4z^3 & 60 \le y^+ < 4.0 & (3.3.12b) \\ 0.30 & y^+ > 60 & (3.3.12c) \end{cases}$$

where $z = \ln y^+$. The constants in Eq. (3.3.12a) are determined by requiring that at $y^+ = 4$,

$$\frac{\tau^+}{k^+} = 0.054, \qquad \left(\frac{\tau^+}{k^+} \right)' = 0.0145 \tag{3.3.13}$$

according to the data of [2].

The constants c_1 to c_4 in Eq. (3.3.12b) are taken as

$$c_1 = 0.080015, \quad c_2 = -0.11169$$
$$c_3 = 0.07821, \quad c_4 = -0.0095665 \tag{3.3.14}$$

3.3.3 Subroutine KEINITG

In this subroutine the rate of dissipation profile ε is determined by assuming

$$(\varepsilon_m)_{\text{CS}} = (\varepsilon_m)_{k\text{-}\varepsilon} = f_\mu c_\mu \frac{k^2}{\varepsilon} \tag{3.3.15}$$

or from

$$\varepsilon = \frac{f_\mu c_\mu k^2}{(\varepsilon_m)_{\text{CS}}}$$

where $(\varepsilon_m)_{\text{CS}}$ is determined from the CS-eddy viscosity model in subroutine EDDY. Whereas f_μ is constant for high Reynolds number flows with a typical value of 1.0, it is not a constant for low Reynolds number flows. Its variation differs according to different models developed close to the wall, say $y^+ \leq 60$ [3].

3.3.4 Subroutine KEWALL

This subroutine provides the wall boundary conditions for the k-ε model which includes low (with wall functions) and high Reynolds number (without wall functions) flows as well as the zonal method. For low Reynolds numbers, there are four physical wall boundary conditions and one "numerical" boundary condition. They are given by Eqs. (2.1.13) and (2.3.3c). The last one is a numerical boundary condition and follows from the definition of u_τ (see Eq. (2.3.1a)).

For high Reynolds numbers, the "wall" boundary conditions are specified at a distance $y_0 = (\nu/u_\tau)y_0^+$. In this case we have a total of five boundary conditions. The physical boundary conditions are given by Eqs. (2.2.5), (2.2.6), and (2.2.7). The numerical boundary condition is given by Eq. (2.4.8).

3.3.5 Subroutine KESOLV

This subroutine performs both forward and backward sweeps for low and high Reynolds numbers, including the zonal method, by using the block elimination method. When the perturbation quantities $um(1, j)$ to $um(8, j)$ are calculated so that new values of f_j, u_j, v_j, etc., are calculated, a relaxation parameter rex is used in order to stabilize the solutions.

In this subroutine, for the zonal method we also reset k, ε in the inner region only. Since the CS model is used for the inner region, there is no need for these quantities. For safety, they are arbitrarily defined in this region.

3.4 Part 4 – Solution Algorithm

When the system of first-order equations to be solved with the block elimination method becomes higher than, say 6, the preparation of the solution algorithm with the recursion formulas described in subroutine SOLV3 becomes tedious. A matrix-solver algorithm (MSA) discussed here can be used to perform the matrix operations required in the block elimination method. This algorithm consists of three subroutines, namely, subroutines GAUSS, GAMSV and USOLV. To illustrate its use, we discuss the replacement of SOLV3 with MSA.

(1) Read in

```
DIMENSION DUMM(3),BB(2,3),YY(3,81),NROW(3,81),GAMJ(2,3,81).
         AA(3,3,81),CC(2,3,81)
DATA IROW,ICOL,ISROW,INP/3,3,3,81/
```

Here IROW, ICOL correspond to number of maximum rows and columns respectively. ISROW denotes the number of "wall" boundary conditions and INP the total number of j-points in the η-direction, and

$$\text{BB} = B_j, \quad \text{YY} = \vec{w}_j, \quad \text{GAMJ} = \Gamma_j, \quad \text{AA} = A_j, \quad \text{CC} = C_j$$

The first and second numbers in the arguments of AA, BB, CC and GAMJ correspond to the number of nonzero rows and columns in A_j (or Δ_j), B_j, C_j and Γ_j matrices, respectively (see subsection 2.3.4, for example). Note that B_j and Γ_j have the same structure and the last row of B_j and the-first two rows of C_j are all zero. The index 81 in YY, NROW, GAMJ, AA and CC refers to INP.

(2) Set the elements of all matrices A_j, B_j, C_j (and Δ_j) equal to zero.

(3) Define the matrices A_0 and C_0 by reading in their elements. Note that only those nonzero elements in the matrices are read in since in (2) we set all the elements equal to zero.

(4) Call subroutine GAUSS.

(5) Read in the elements of B_j and call subroutine GAMSV to compute Γ_1.

(6) Define A_j according to Eq. (2.3.23b), call GAUSS and read in the elements of C_j.

(7) Recall the elements of B_j and call GAMSV to compute Γ_2.

(8) Repeat (6) and (7) for $j < J$.

(9) At $j = J$, read in the last row of A_J which is also equal to the last row of Δ_J.

(10) Compute $\vec{w}_0$ according to Eq. (2.3.31a). Here $\vec{r}_0 = \text{RRR}(1,81)$.

(11) Define the right-hand side of Eq. (2.3.31b) and compute $\vec{w}_j$ according to Eq. (2.3.31b).

(12) In the backward sweep, with δ_j corresponding to UM(I,J), compute $\vec{\delta}_J$ according to Eq. (2.3.32a) by calling USOLV at INP.

(13) Define the right-hand side of Eq. (2.3.32b) and solve for δ_j by calling USOLV for $j = J - 1, J - 2, \ldots, 0$.

This algorithm is very useful to solve the linear system for the k-ε model equations. With all A_j, B_j, C_j matrices and $\overline{\overline{r}}_j$ nicely defined in subroutine KECOEF, the solution of Eq. (2.3.15) is relatively easy.

3.5 Part 5 – Basic Tools

This part of the computer program includes basic tools to perform smoothing, differentiation, integration and interpolation. For example, subroutine DIFF-3 provides first, second and third derivatives of the input function at inputs. First derivatives use weighted angles, second and third derivatives use cubic fits. Subroutine INTRP3 provides cubic interpolation. Given the values of a function (F1) and its derivatives at N1 values of the independent variable (X1), this subroutine determines the values of the function (F2) at N2 values of the independent variable (X2). Here X2 can be in arbitrary order.

Another subroutine used for interpolation is subroutine LNTP; it performs linear interpolation.

References

[1] Cebeci, T. and Cousteix, J.: Modeling and Computation of Boundary-Layer Flows, Horizons Publishing, Long Beach, CA and Springer, Heidelberg, Germany, 1998.
[2] Mansour, N. N., Kim, J., and Moin, P.: Near-Wall k-ε Turbulence Modeling. *J. AIAA*, **27**, 1068–1073, 1989.
[3] Cebeci, T.: Analysis of Turbulent Flows. Elsevier, 2003.

4 Computer Program for CS Model for Flows with Separation

4.0 Introduction

As discussed, for example, in [1], for a given external velocity distribution, the solutions of the boundary-layer equations are singular at separation. To continue calculations beyond separation, it is necessary to compute the external velocity or pressure as part of the solution. This procedure is known as the *inverse* mode. Catherall and Mangler [2] were the first to show that modification of the external velocity distribution near the region of flow separation leads to solutions free of numerical difficulties. By prescribing the displacement thickness or wall shear stress as a boundary condition in addition to the usual boundary conditions, the boundary-layer equations can be integrated through the separation location and into the region of reverse flow without any evidence of singularity at the separation point [1].

A problem associated with using an inverse mode to solve the boundary-layer equations is the lack of a priori knowledge of the required displacement thickness or the wall shear. The appropriate value must be obtained as part of the overall problem from the interaction between the boundary layer and the inviscid flow. In the case of internal flow, the problem is somewhat easier because the conservation of mass in integral form can be used to relate pressure $p(x)$ to velocity $u(x, y)$ in terms of mass balance in the duct.

For two-dimensional external flows, two procedures have been developed to couple the solutions of the inviscid and viscous equations. In the first procedure, developed by LeBalleur [3] and Carter and Wornom [4], the solution of the boundary-layer equations is obtained in the standard mode, and a displacement-thickness, $\delta^{*^{\circ}}$, distribution is determined. If this initial calculation encounters separation, then $\delta^{*^{\circ}}(x)$ is extrapolated to the trailing edge, and one complete cycle of the viscous and inviscid calculation is performed with the boundary-layer equations now solved in the inverse mode. This will, in general, lead to two different external velocity distributions, $u_{ev}(x)$ derived from the inverse boundary-layer solution, and $u_{ei}(x)$ derived from the updated approximation to the inviscid velocity past the airfoil with the added displacement thickness. A relaxation formula is introduced to define an updated displacement-thickness distribution,

$$\delta^* = \delta^{*^\circ}(x)\left\{1 + \omega\left[\frac{u_{ev}(x)}{u_{ei}(x)} - 1\right]\right\} \tag{4.0.1}$$

where ω is a relaxation parameter, and the procedure is repeated with this updated mass flux.

In the second approach, developed by Veldman [5], the external velocity $u_e(x)$ and the displacement thickness $\delta^*(x)$ are treated as unknown quantities, and the equations are solved in an inverse mode simultaneously in successive sweeps over the airfoil surface. For each sweep, the external boundary condition for the boundary-layer equation in dimensionless form, with $u_e(x)$ normalized with u_∞, is written as

$$u_e(x) = u_e^0(x) + \delta u_e(x) \tag{4.0.2a}$$

where $u_e^0(x)$ denotes the inviscid velocity and δu_e the perturbation due to the displacement thickness, which is calculated from the Hilbert integral

$$\delta u_e(x) = \frac{1}{\pi}\int_{x_a}^{x_b}\frac{d}{d\sigma}(u_e\delta^*)\frac{d\sigma}{x - \sigma} \tag{4.0.2b}$$

The above expression is based on the thin airfoil approximation with the term $d/d\sigma(u_e\delta^*)$ denoting the blowing velocity used to simulate the boundary layer in the interaction region (x_a, x_b). This approach is more general and will be used in the present approach for calculating boundary-layer flows with separation.

We consider a laminar and turbulent flow. We assume the calculations start at the leading edge, $x = 0$, for laminar flow and are performed for turbulent flow at any x-location by specifying the transition location. The use of the two-point finite-difference approximations for streamwise derivatives is proper and does not cause numerical difficulties if there is no flow separation. If there is one, then it is necessary to use backward difference formulas as discussed in subsection 2.3.1.

We employ two separate but closely related transformations. The first one is the Falkner-Skan transformation in which the dimensionless similarity variable η and a dimensionless stream function $f(x, \eta)$ are defined by Eqs. (2.1.1). It is a convenient and useful transformation to generate initial conditions which start either as a flat plate flow or stagnation flow. It is also useful to perform the boundary-layer calculations subject to the boundary conditions that correspond to the standard mode (external velocity distribution given). The resulting equations from this transformation are given by Eqs. (2.1.3).

In the inverse mode, since $u_e(x)$ is also an unknown, slight changes are made to the transformation given by Eq. (2.1.1), replacing $u_e(x)$ by u_∞ and redefining new variables Y and F by

$$Y = \sqrt{u_\infty/\nu x}\, y, \quad \psi(x, y) = \sqrt{u_\infty \nu x}\, F(\xi, Y), \quad \xi = \frac{x}{L} \tag{4.0.3}$$

The resulting equation and its wall boundary equations can be written as

$$(bF'')' + \frac{1}{2}FF'' = \xi \left(F'\frac{\partial F'}{\partial \xi} - F''\frac{\partial F}{\partial \xi} \right) - \xi w \frac{dw}{d\xi} \tag{4.0.4}$$

$$Y = 0, \quad F' = 0, \quad F = 0 \tag{4.0.5}$$

Here primes denote differentiation with respect to Y and $w = u_e/u_\infty$.

The edge boundary condition is obtained from Eq. (4.0.2). By applying a discretization approximation to the Hilbert integral, Eq. (4.0.2b), we can write (see subsection 4.2.3)

$$u_e(x_i) = u_e^0(\xi_i) + C_{ii}D_i + \sum_{j=1}^{i-1} C_{ij}D_j + \sum_{j=i+1}^{N} C_{ij}D_j \tag{4.0.6}$$

where the subscript i denotes the ξ-station where the inverse calculations are to be performed, C_{ij} is a matrix of interaction coefficients obtained by the procedure described in subroutine HIC, and D is given by $D = u_e\delta^*$. In terms of transformed variables, the parameter D becomes

$$\overline{D} = \frac{D}{Lu_\infty} = \sqrt{\frac{\xi}{R_L}}\,(Y_e w - F_e) \tag{4.0.7}$$

and the relation between the external velocity u_e and displacement thickness δ^* provided by the Hilbert integral can then be written in dimensionless form as

$$Y = Y_e, \quad F_e'(\xi^i) - \lambda[Y_e F_e'(\xi^i) - F_e(\xi^i)] = g_i \tag{4.0.8}$$

where

$$\lambda = C_{ii}\sqrt{\xi^i/R_L} \tag{4.0.9a}$$

$$g_i = \overline{u_e^0}(\xi^i) + \sum_{j=1}^{i-1} C_{ij}\overline{D}_j + \sum_{j=i+1}^{N} C_{ij}\overline{D}_j \tag{4.0.9b}$$

4.1 Numerical Method

The numerical method for the inverse problem is similar to the numerical method described for the standard problem in Section 2.3. Since $u_e(\xi)$ must be computed as part of the solution procedure, we treat it as an unknown. Remembering that the external velocity w is a function of ξ only, we write

$$w' = 0 \tag{4.1.1}$$

As in the case of the standard problem, new variables $U(\xi, Y)$, $V(\xi, Y)$ are introduced and Eq. (4.0.4) and its boundary conditions, Eq. (4.0.5) and (4.0.8), are expressed as a first-order system,

$$F' = U \tag{4.1.2a}$$

$$U' = V \tag{4.1.2b}$$

$$(bV)' + \frac{1}{2}FV = \xi\left(U\frac{\partial U}{\partial \xi} - V\frac{\partial F}{\partial \xi}\right) - \xi w\frac{dw}{d\xi} \tag{4.1.2c}$$

$$Y = 0, \qquad F = U = 0 \tag{4.1.3a}$$

$$Y = Y_e, \qquad U = w, \qquad \lambda F + (1 - \lambda Y_e)w = g_i \tag{4.1.3b}$$

Finite-difference approximations to Eqs. (4.1.1) and (4.1.2) are written in a similar fashion to those expressed in the original Falkner-Skan variables, yielding

$$h_j^{-1}(w_j^n - w_{j-1}^n) = 0 \tag{4.1.4a}$$

$$h_j^{-1}(F_j^n - F_{j-1}^n) = U_{j-1/2}^n \tag{4.1.4b}$$

$$h_j^{-1}(U_j^n - U_{j-1}^n) = V_{j-1/2}^n \tag{4.1.4c}$$

$$h_j^{-1}(b_j^n V_j^n - b_j^n V_{j-1}^n) + \left(\frac{1}{2} + \alpha^n\right)(FV)_{j-1/2}^n$$
$$+ \alpha^n[(w^2)_{j-1/2}^n - FLARE(U^2)_{j-1/2}^n]$$
$$+ \alpha^n(V_{j-1/2}^{n-1}F_{j-1/2}^n - F_{j-1/2}^{n-1}V_{j-1/2}^n) = R_{j-1/2}^{n-1} \tag{4.1.4d}$$

where

$$R_{j-1/2}^{n-1} = -L_{j-1/2}^{n-1} + \alpha^n[(FV)_{j-1/2}^{n-1} - FLARE(U^2)_{j-1/2}^{n-1}] \tag{4.1.5a}$$

$$L_{j-1/2}^{n-1} = [h_j^{-1}(b_j v_j - b_{j-1}v_{j-1}) + \frac{1}{2}(FV)_{j-1/2} - \alpha^n(w^2)_{j-1/2}]^{n-1} \tag{4.1.5b}$$

In Eq. (4.1.4d), the parameter $FLARE$ refers to the Flügge-Lotz-Reyhner approximation [1] used to set $u\frac{\partial u}{\partial x}$ equal to zero in the momentum equation wherever $u < 0$. As a result, the numerical instabilities that plague attempts to integrate the boundary-layer equations against the local direction of flow are avoided. In regions of positive streamwise velocity $(u_j > 0)$, it is taken as unity and as zero in regions of negative streamwise velocity $(u_j \leq 0)$.

The linearized form of Eqs. (4.1.4) and (4.1.3) can be expressed in the form given by Eq. (2.4.15) or

$$
\begin{array}{cccccccccccc}
\delta F_0 & \delta U_0 & \delta V_0 & \delta w_0 & \delta F_j & \delta U_j & \delta V_j & \delta w_j & \delta F_J & \delta U_J & \delta V_J & \delta w_J
\end{array}
$$

$$
\text{b.c.}\;\left[
\begin{array}{cccc:cccc:cccc}
1 & 0 & 0 & 0 & 0 & 0 & 0 & 0 & & & & \\
0 & 1 & 0 & 0 & 0 & 0 & 0 & 0 & & & & \\
0 & -1 & \frac{-h_1}{2} & 0 & 0 & 1 & \frac{-h_1}{2} & 0 & & & & \\
0 & 0 & 0 & -1 & 0 & 0 & 0 & 1 & & & & \\
\hdashline
-1 & \frac{-h_j}{2} & 0 & 0 & 1 & \frac{-h_j}{2} & 0 & 0 & 0 & 0 & 0 & 0 \\
(s_4)_j & (s_6)_j & (s_2)_j & (s_8)_j & (s_3)_j & (s_5)_j & (s_1)_j & (s_7)_j & 0 & 0 & 0 & 0 \\
0 & 0 & 0 & 0 & 0 & -1 & \frac{-h_{j+1}}{2} & 0 & 0 & 1 & \frac{-h_{j+1}}{2} & 0 \\
0 & 0 & 0 & 0 & 0 & 0 & 0 & -1 & 0 & 0 & 0 & 1 \\
\hdashline
& & & & -1 & \frac{-h_j}{2} & 0 & 0 & 1 & \frac{-h_j}{2} & 0 & 0 \\
& & & & (s_4)_J & (s_6)_J & (s_2)_J & (s_8)_J & (s_3)_J & (s_5)_J & (s_1)_J & (s_7)_J \\
& & & & 0 & 0 & 0 & 0 & \gamma_1 & 0 & 0 & \gamma_2 \\
& & & & 0 & 0 & 0 & 0 & 0 & 1 & 0 & -1
\end{array}
\right]
\begin{array}{c}
\\ \\ \\ \\ \text{b.c.} \\ \text{b.c.}
\end{array}
$$

$$
\begin{bmatrix}
\delta F_0 \\ \delta U_0 \\ \delta V_0 \\ \delta w_0 \\
\delta F_j \\ \delta U_j \\ \delta V_j \\ \delta w_j \\
\delta F_J \\ \delta U_J \\ \delta V_J \\ \delta w_J
\end{bmatrix}
=
\begin{bmatrix}
(r_1)_0 \\ (r_2)_0 \\ (r_3)_0 \\ (r_4)_0 \\
(r_1)_j \\ (r_2)_j \\ (r_3)_j \\ (r_4)_j \\
(r_1)_J \\ (r_2)_J \\ (r_3)_J \\ (r_4)_J
\end{bmatrix}
\tag{4.1.6}
$$

with δ_j and r_j now defined by

$$
\delta_j = \begin{vmatrix} \delta F_j \\ \delta U_j \\ \delta V_j \\ \delta w_j \end{vmatrix}, \qquad
r_j = \begin{vmatrix} (r_1)_j \\ (r_2)_j \\ (r_3)_j \\ (r_4)_j \end{vmatrix}
\tag{4.1.7}
$$

and A_j, B_j, C_j becoming 4×4 matrices defined by

$$
A_0 = \begin{vmatrix}
1 & 0 & 0 & 0 \\
0 & 1 & 0 & 0 \\
0 & -1 & -\frac{h_1}{2} & 0 \\
0 & 0 & 0 & -1
\end{vmatrix}, \qquad
A_j = \begin{vmatrix}
1 & -\frac{h_j}{2} & 0 & 0 \\
(s_3)_j & (s_5)_j & (s_1)_j & (s_7)_j \\
0 & -1 & -\frac{h_{j+1}}{2} & 0 \\
0 & 0 & 0 & -1
\end{vmatrix}, \qquad 1 \leq j \leq J-1
\tag{4.1.8a}
$$

$$
A_J = \begin{vmatrix}
1 & -\frac{h_J}{2} & 0 & 0 \\
(s_3)_J & (s_5)_J & (s_1)_J & (s_7)_J \\
\gamma_1 & 0 & 0 & \gamma_2 \\
0 & 1 & 0 & -1
\end{vmatrix}, \qquad
B_j = \begin{vmatrix}
-1 & -\frac{h_j}{2} & 0 & 0 \\
(s_4)_j & (s_6)_j & (s_2)_j & (s_8)_j \\
0 & 0 & 0 & 0 \\
0 & 0 & 0 & 0
\end{vmatrix}, \qquad 1 \leq j \leq J
\tag{4.1.8b}
$$

$$
C_j = \begin{vmatrix}
0 & 0 & 0 & 0 \\
0 & 0 & 0 & 0 \\
0 & 1 & -\frac{h_{j+1}}{2} & 0 \\
0 & 0 & 0 & 1
\end{vmatrix}, \qquad 0 \leq j \leq J-1
\tag{4.1.8c}
$$

Here the first two rows of A_0 and C_0 and the last two rows of B_J and A_J correspond to the linearized boundary conditions,

$$\delta F_0 = \delta U_0 = 0; \quad \delta U_J - \delta w_J = w_J - U_J, \quad \gamma_1 \delta F_J + \gamma_2 \delta w_J = \gamma_3 \qquad (4.1.9)$$

where

$$\gamma_1 = \lambda, \qquad \gamma_2 = 1 - \lambda Y_J, \qquad \gamma_3 = g_i - (\gamma_1 F_J + \gamma_2 w_J) \qquad (4.1.10)$$

As a result

$$(r_1)_0 = (r_2)_0 = 0 \qquad (4.1.11a)$$

$$(r_3)_J = \gamma_3, \qquad (r_4)_J = w_J - U_J \qquad (4.1.11b)$$

The third and fourth rows of A_0 and C_0 correspond to Eq. (3.2.1b) and the linearized form of Eq. (2.3.6a), that is,

$$\delta w_j - \delta w_{j-1} = w_{j-1} - w_j = (r_4)_{j-1} \qquad (4.1.12)$$

if the unknowns f, u, v are replaced by F, U and V. Similarly, the first and second rows of A_j and B_j correspond to Eq. (3.2.1a) and (2.3.6c) with two terms added to its left-hand side,

$$(s_7)_j \delta w_j + (s_8)_j \delta w_{j-1} \qquad (4.1.13a)$$

with $(s_7)_j$ and $(s_8)_j$ defined by

$$(s_7)_j = \alpha^n w_j, \qquad (s_8)_j = \alpha^n w_{j-1} \qquad (4.1.13b)$$

The coefficients $(s_1)_j$ to $(s_6)_j$ defined by Eqs. (3.2.5) remain unchanged provided we set

$$\alpha_1 = \frac{1}{2} + \alpha^n, \qquad \alpha_2 = \alpha^n \qquad (4.1.14)$$

and define $(r_2)_j$ by

$$\begin{aligned}
(r_2)_j = {} & R_{j-1/2}^{n-1} - [h_j^{-1}(b_j V_j - b_{j-1} V_{j-1}) + (\tfrac{1}{2} + \alpha^n)(FV)_{j-1/2} \\
& + \alpha^n[(w^2)_{j-1/2} - FLARE(U^2)_{j-1/2}] \\
& + \alpha^n(V_{j-1/2}^{n-1} F_{j-1/2} - F_{j-1/2}^{n-1} V_{j-1/2})]
\end{aligned} \qquad (4.1.15)$$

The remaining elements of the r_j vector follow from Eqs. (3.2.2), (4.1.12) and (4.1.15) so that, for $l \le j \le J$, $(r_1)_j$, $(r_2)_j$, $(r_3)_{j-1}$ are given by Eqs. (3.2.2a), (4.1.15) and (3.2.2b), respectively. For the same j-values, $(r_4)_{j-1}$ is given by the right-hand side of Eq. (4.1.12).

The parameters γ_1, γ_2 and γ_3 in Eq. (4.1.9) determine whether the system given by the linearized form of Eqs. (4.1.4) and their boundary conditions is to be solved in standard or inverse form. For an inverse problem, they are represented by the expressions given in Eq. (4.1.9) and for a standard problem by $\gamma_1 = 0$, $\gamma_2 = 1.0$ and $\gamma_3 = 0$.

It should be noted that for flows with separation, it is necessary to use backward differences as discussed for the CS and k-ε models in subsection 2.3.1.

In that case, the coefficients $(s_1)_j$ to $(s_6)_j$ are given by Eq. (2.3.13), and $(r_2)_j$ by Eq. (2.3.14) with the relations given by Eq. (4.1.15).

The solution of Eq. (2.3.15), with $\underline{\delta}_j$ and $\underline{r}_j$ defined by Eq. (4.1.7) and with A_j, B_j and C_j matrices given by Eqs. (4.1.8), can again be obtained by the block-elimination of subsection 2.3.4. The resulting algorithm, similar to SOLV3, called SOLV4, is given in the accompanying CD-ROM.

Numerical Method for Wake Flows

In interaction problems involving airfoils, it is usually sufficient to neglect the wake effect and perform calculations on the airfoil only, provided that there is no or little flow separation on the airfoil. With flow separation, the relative importance of including the wake effect in the calculations depends on the flow separation as shown in Fig. 4.1 taken from [1]. Figure 4.1a shows the computed separation locations on a NACA 0012 airfoil at a chord Reynolds number, R_c of 3×10^6. When the wake effect is included, separation is encountered for angles of attack greater than $10°$, and attempts to obtain results without consideration of the wake effect lead to erroneously large regions of recirculation that increases with angle of attack, as discussed in [1]. Figure 4.1b shows that the difference in displacement thickness at the trailing edge is negligible for $\alpha = 10°$ but more than 30% for $\alpha = 16°$.

As discussed in [1], the inverse boundary layer method described here can also be extended to include wake flows. This requires the specification of a turbulence model for wake flows and minor modifications to the numerical method.

The extension of the CS model for wall boundary layers to wake flows is given by the following expressions described in [1]:

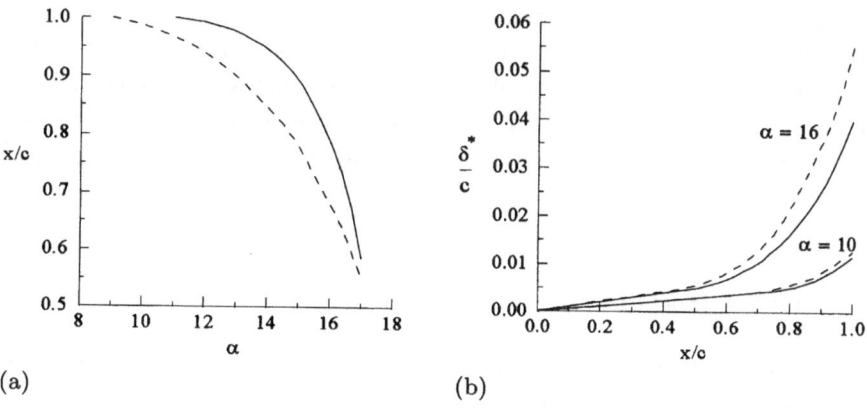

(a)   (b)

Fig. 4.1. Wake effect on (**a**) flow separation and (**b**) displacement thickness – NACA 0012 airfoil. —, with wake; – – –, without wake.

$$\varepsilon_m = (\varepsilon_m)_w + [(\varepsilon_m)_{\text{t.e.}} - (\varepsilon_m)_w] \exp \frac{-(x - x_{\text{t.e.}})}{\lambda \delta_{\text{t.e.}}} \tag{4.1.16}$$

where $\delta_{\text{t.e.}}$ is the boundary layer thickness at the trailing edge, λ is an empirical parameter, $(\varepsilon_m)_{\text{t.e.}}$ is the eddy viscosity at the trailing edge, and $(\varepsilon_m)_w$ is the eddy-viscosity in the far wake given by the larger of

$$(\varepsilon_m)_w^l = 0.064 \int_{-\infty}^{y_{\min}} (u_e - u)\, dy \tag{4.1.17}$$

and

$$(\varepsilon_m)_w^u = 0.064 \int_{y_{\min}}^{\infty} (u_e - u)\, dy \tag{4.1.18}$$

with $y_{\min}$ denoting the location where the velocity is a minimum.

The studies conducted in [6] indicate that a choice of $\lambda = 20$ is satisfactory for single airfoils. Calculations with different values of λ essentially produced similar results, indicating that the modeling of wake flows with Eq. (4.1.16) was not too sensitive to the choice of λ. The application of the above model to wake flows with strong adverse pressure gradient, however, indicated that this was not the case and the value of the parameter is an important one. On the basis of that study, a value of $\lambda = 50$ was found to produce best results and is used in the present computer program.

A modification to the numerical method of the previous section arises due to the boundary conditions along the wake dividing streamline. The new "wall" boundary conditions on f and u now become

$$\eta = 0, \qquad f_0 = 0, \qquad v_0 = 0 \tag{4.1.19}$$

so that the second row of Eq. (4.1.6) can de written as

$$0 \quad 0 \quad 1 \quad 0 \tag{4.1.20}$$

Before the boundary-layer equations can be solved for wake flows, the initial velocity profiles must satisfy the wall and edge boundary conditions. When the calculations are first performed for wall boundary layer flows and are then to be extended to wake flows, it is necessary to modify the velocity profiles computed for wall boundary layers. This is done in subroutine WAKEPR of the computer program.

4.2 Computer Program

In the accompanying CD-ROM, we present a computer program for obtaining boundary-layer solutions on airfoils and their wakes. The code can be used to solve both standard and inverse problems. For standard problems, it is similar to the computer program described in Section 3.2 except that it also provides solutions to wake flows and is applicable to both laminar and turbulent flows. For inverse problems, it uses the numerical procedure described in the previous sections, and when coupled to an inviscid flow, it can also be used in inviscid-viscous interaction problems, as described in Section 4.3.

The computer program consists of a MAIN and 15 subroutines, INPUT, IVPL, HIC, EDDY, SWTCH, COEF, WAKEPR, DIFF1, LNTP, INTEG, AMEAN, SOLV4, EDGCHK, CALFA and GAMCAL. MAIN, as before, is used to control the logic of the computations. Here the parameter g_i in Eq. (4.0.8) is also calculated, with

$$\text{SUM1} = \sum_{j=1}^{i-1} C_{ij}\overline{D}_j \tag{4.2.1a}$$

and

$$\text{SUM2} = \sum_{j=i+1}^{N} C_{ij}\overline{D}_j \tag{4.2.1b}$$

The initial displacement thickness (δ^*) distribution needed in the calculation of $\overline{D}_j$ is computed in subroutine INPUT by assuming a δ^* distribution based on a flat-plate flow and given by

$$\frac{\delta^*}{x} = 0.036\, H R_x^{-0.20} \tag{4.2.2}$$

with $H = 1.3$.

Of the 15 subroutines, subroutine WAKEPR is used to modify the profiles resulting from wall boundary layers for wake profiles. Except for this subroutine and except for subroutines INPUT, IVPL and HIC, the remaining subroutines are similar to those described in Sections 3.2 and 3.3. For this reason, only these three subroutines are described below.

4.2.1 Subroutine INPUT

This subroutine is used to generate the grid, calculate γ_{tr} in the eddy viscosity formulas, initial δ^*-distribution, and pressure gradient parameters m and m_1. The following data are read in and the number of j-points J(NP) is computed from Eq. (3.1.11).

NXT	Total number of x-stations
NXTE	Total number of x-stations on the body
NXS	NX-station after which inverse calculations begin
NPT	Total number of η-grid points
ISWPT	Number of sweeps in the inverse problem. Usually a value of 10 is sufficient for low angles of attack; higher values are needed for high angles of attack

RL   Reynolds number, $u_\infty c/\nu$
XTR x/c value for transition location
ETAE   Transformed boundary layer thickness η_e at $x = 0$, ETAE = 8.0
VGP K is the variable-grid parameter. Take $K = 1.0$ for
            laminar flow and $K = 1.14$ for turbulent flow. For a flow
            consisting of both laminar and turbulent regions, take $K = 1.14$
DETA(1) $\Delta\eta(h_1)$ – initial step size of the variable grid system.
            Take $h_1 = 0.01$ for turbulent flows
P2(1) m at $x = 0$ (NX = 1)
$x/c, y/c$ Dimensionless airfoil coordinates
u_e/u_∞ Dimensionless external velocity

4.2.2 Subroutine IVPL

At $x = 0$, Eq. (2.1.3) reduces to the Falkner-Skan equation, which can be solved subject to the boundary conditions of Eq. (2.1.13a) and (2.1.14a). Since the equations are solved in linearized form, initial estimates of f_j, u_j and v_j are needed in order to obtain the solutions of the nonlinear Falkner-Skan equation. Various expressions can be used for this purpose. Since Newton's method is used, however, it is useful to provide as good an estimate as is possible and an expression of the form.

$$u_j = \frac{3}{2}\frac{\eta_j}{\eta_e} - \frac{1}{2}\left(\frac{\eta_j}{\eta_e}\right)^3 \tag{4.2.3}$$

usually satisfies this requirement. The above equation is obtained by assuming a third-order polynomial and by determining constants a, b, c from the boundary conditions from one of the properties of momentum equation which requires that $f'' = 0$ at $\eta = \eta_e$.

The other profiles f_j, v_j follow from Eq. (4.2.3) and are obtained by integrating and/or differentiating Eq. (4.2.3) for f_j and v_j, respectively.

4.2.3 Subroutine HIC

This subroutine calculates the coefficients of the Hilbert integral denoted by C_{ij}. While they can be generated from any suitable integration procedure, we use the following procedure which is appropriate with the box method [1].

We evaluate

$$H_i = \int_{\xi^i}^{\xi^L} G(\sigma)\frac{d\sigma}{\xi^i - \sigma} \tag{4.2.4}$$

where

$$G(\sigma) = \frac{dF}{d\sigma}$$

with F denoting any function, so that over each subinterval (ξ^{n-1}, ξ^n), except the two enclosing the point $\xi = \xi^i$, we replace $G(\sigma)$ by its midpoint value:

$$\int_{\xi^{n-1}}^{\xi^n} \frac{G(s)\,d\sigma}{\xi^i - \sigma} = G_{n-1/2} \int_{\xi^{n-1}}^{\xi^n} \frac{d\sigma}{\xi^i - \sigma} = G_{n-1/2} \ln \left| \frac{\xi^i - \xi^{n-1}}{\xi^i - \xi^n} \right| \qquad (4.2.5)$$

Making the further approximation,

$$G_{n-1/2} = \frac{F_n - F_{n-1}}{\xi^n - \xi^{n-1}}$$

we can write

$$\int_{\xi^{n-1}}^{\xi^n} \frac{dF}{d\sigma} \frac{d\sigma}{\xi^i - \sigma} = E_n^i (F_n - F_{n-1}) \qquad (4.2.6)$$

where for $n \neq i$ or $i+1$

$$E_n^i = (\xi^n - \xi^{n-1})^{-1} \ln \left| \frac{\xi^i - \xi^{n-1}}{\xi^i - \xi^n} \right| \qquad (4.2.7)$$

for the two subintervals ξ^{i-1} to ξ^i and ξ^i to ξ^{i+1}. Because of the cancellation with the constant term, account should be taken of the linear variation of G from one interval to the next. Thus, we take the linear interpolation

$$G = \frac{G_{i-1/2}(\xi^{i+1} - \xi^i) + G_{i+1/2}(\xi^i - \xi^{i-1}) + 2(G_{i+1/2} - G_{i-1/2})(\sigma - \xi^i)}{\xi^{i+1} - \xi^{i-1}}$$

so that

$$\int_{\xi^{i-1}}^{\xi^{i+1}} \frac{G\,d\sigma}{\xi^i - \sigma} = \frac{G_{i-1/2}(\xi^{i+1} - \xi^i) + G_{i+1/2}(\xi^i - \xi^{i-1})}{\xi^{i+1} - \xi^{i-1}} \ln \left| \frac{\xi^i - \xi^{i-1}}{\xi^i - \xi^{i+1}} \right| \qquad (4.2.8)$$
$$- 2(G_{i+1/2} - G_{i-1/2})$$

Replacing the midpoint derivative values by difference quotients, we obtain

$$\int_{\xi^{i-1}}^{\xi^{i+1}} \frac{dF}{d\sigma} \frac{d\sigma}{\xi^i - \sigma} = E_i^i (F_i - F_{i-1}) + E_{i+1}^i (F_{i+1} - F_i) \qquad (4.2.9)$$

where

$$E_i^i = \frac{\dfrac{\xi^{i+1} - \xi^i}{\xi^{i+1} - \xi^{i-1}} \ln \left| \dfrac{\xi^i - \xi^{i-1}}{\xi^i - \xi^{i+1}} \right| + 2}{\xi^i - \xi^{i-1}} \qquad (4.2.10a)$$

$$E_{i+1}^i = \frac{\dfrac{\xi^i - \xi^{i-1}}{\xi^{i+1} - \xi^{i-1}} \ln \left| \dfrac{\xi^i - \xi^{i-1}}{\xi^i - \xi^{i+1}} \right| - 2}{\xi^{i+1} - \xi^i} \qquad (4.2.10b)$$

Thus

$$H_i = E_2^i(F_2 - F_1) + E_3^i(F_3 - F_2) + \ldots + E_{L-1}^i(F_{L-1} - F_{L-2}) + E_L^i(F_L - F)$$
$$= -E_2^i F_1 + (E_2^i - E_3^i)F_2 + \ldots + (E_{L-1}^i - E_L^i)F_{L-1} + E_L^i F_L$$

$$(4.2.11)$$

so, finally the C_{ij} of Eq. (4.0.6) are given by

$$C_{ij} = \frac{1}{\pi}(E_j^i - E_{j+1}^i) \tag{4.2.12}$$

and the E^i are given by Eqs. (4.2.10a) and (4.2.10b) with $E_1^i = E_{L+1}^i = 0$.

References

[1] Cebeci, T., An Engineering Approach to the Calculation of Aerodynamic Flows. Horizons Publishing, Long Beach, CA and Springer, Heidelberg, Germany, 1999.

[2] Catherall, D. and Mangler K. W., The Integration of the Two-Dimensional Laminar Boundary-Layer Equations Past the Point of Vanishing Skin Friction. *J. Fluid Mechanics*, **26**, 163, 1966.

[3] LeBalleur, J. C., "Couplage Visqueux-non Visqueux: Analyse du Probleme Incluant Decollements et Ondes de Choc," La Rech. Aerosp., 1977-6, 349. English translation in ESA TT476, 1977.

[4] Carter, J. and Wornom, S. F., "Solutions for Incompressible Separated Boundary Layers Including Viscous-Inviscid Interaction," in Aerodynamic Analysis Requiring Advanced Computers, NASA SP-347, p. 125, 1975.

[5] Veldman, A. E. P., New Quasi-Simultaneous Method to Calculate Interacting Boundary Layers. *AIAA J.*, **19**, 769, 1981.

[6] Cebeci, T. and Chang, K. C., "Compressibility and Wake Effects on the Calculations of Flow over High Lift Configuration," 48th Annual Conference, Canadian Aeronautics and Space Institute, Toronto, Canada, April 2001.

5 Hess-Smith Panel Method with Viscous Effects

5.0 Introduction

Calculation of separating flows using boundary-layer theory requires inviscid and boundary-layer methods in which the calculations are performed in an iterative manner so that each successive inviscid flow solution provides the pressure distribution for the next boundary layer solution. The displacement (viscous) effect is then used to modify the inner boundary conditions for a near inviscid calculation; the procedure in this "cycle" known as the interactive boundary-layer scheme must then be repeated until convergence is obtained, see Fig. 5.1.

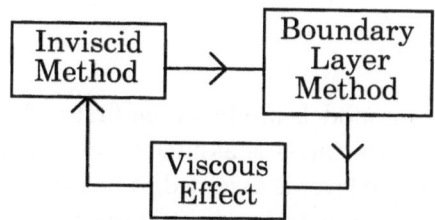

Fig. 5.1. Interactive boundary-layer scheme.

For incompressible flows, a panel method is an ideal inviscid method for this approach. Of the several panel methods, here we choose the one due to Hess and Smith [1]. After a brief description of the mathematical models for the interaction process in an inviscid flow (Section 5.1), we discuss the Hess-Smith (HS) panel method (Section 5.2). The procedure for incorporating the viscous effects into the panel method is discussed in Section 5.3. Changes required in an inviscid method to extend the viscous flow calculations into the wake of an

airfoil are discussed in Section 5.4. A brief description of the computer program
for the HS method with viscous effects is given in Section 5.5.

5.1 Mathematical Models for the Interaction Process in Inviscid Flows

Viscous effects can be incorporated into an inviscid method by replacing the real
flow with an equivalent fictitious inviscid flow such that the velocity components
at the edge of the boundary layer are equal in both cases. We assume that in the
fictitious flow, $u = u_e$ for $y \leq \delta$. This fictitious flow can be modeled by using
either the "solid" displacement surface concept or the "transpiration" model
concept proposed by Lighthill [2]. Figure 5.2a shows the real flow and Figs.
5.2b and 5.2c the equivalent fictitious flow using the displacement surface and
blowing velocity concepts, respectively for two-dimensional flows.

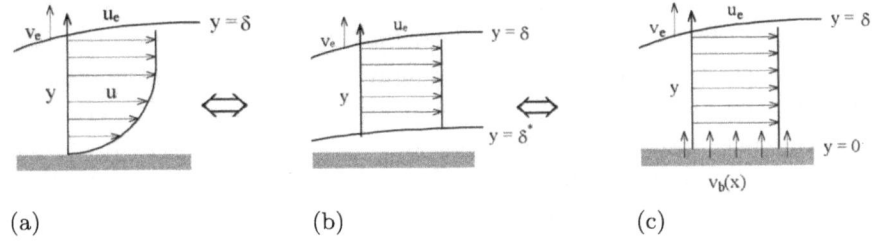

(a)   (b)   (c)

Fig. 5.2. Model of viscous effects in inviscid method in two dimensions (a) real flow, (b),
(c) equivalent fictitious flow.

Both displacement surface and transpiration models are based on first-order
boundary-layer theory, that is; the shear layer thickness is assumed to be thin
and the transverse variation in pressure across the layer is assumed to be negligible. When this is not the case, it is necessary to use second-order boundary-layer
theory, as discussed in [3].

To describe the mathematical models for each concept, it is convenient, following the discussion in [3], to work in terms of the difference between the two
flows corresponding to the real viscous flow and an equivalent inviscid flow defined as analytical continuation, or smooth extrapolation, of the inviscid flow
outside the shear layers in the direction normal to the interface between them.
For a two-dimensional compressible flow with ϱu representing the real viscous
flow and $\varrho_i u_i$ the equivalent inviscid flow the difference of the continuity equations yields

$$\frac{\partial}{\partial x}(\varrho_i u_i - \varrho u) + \frac{\partial}{\partial y}(\varrho_i v_i - \varrho v) = 0 \tag{5.1.1}$$

Integrating across the boundary layer from $y = 0$ to δ and representing the values at the wall by subscript w, we get

$$\varrho_{iw} v_{iw} = \frac{d}{dx} \int_0^\delta (\varrho_i u_i - \varrho u) dy \tag{5.1.2}$$

since both $(\varrho_i v_i - \varrho v)$ and $(\varrho_i u_i - \varrho u)$ vanish at $y = \delta$ because of the matching condition, while $v = 0$ at the wall $(y = 0)$. If a displacement thickness, δ_A^* is defined by

$$\delta_A^* = \frac{1}{\varrho_{iw} u_{iw}} \int_0^\delta (\varrho_i u_i - \varrho u) dy \tag{5.1.3}$$

then we obtain the blowing velocity defined by Lighthill,

$$v_{iw} = \frac{1}{\varrho_{iw}} \frac{d}{dx} (\varrho_{iw} u_{iw} \delta_A^*) \tag{5.1.4}$$

The "classical" displacement thickness, δ_B^*, is defined by the condition that the total mass flow in the equivalent inviscid flow shall equal that in the real viscous flow, so that

$$\int_{\delta_B^*}^\delta \varrho_i u_i dy = \int_0^\delta \varrho u dy \tag{5.1.5}$$

It is shown in [3] that $y = \delta_B^*$ is indeed a streamline of the equivalent inviscid flow.

The relation between these two alternative definitions of displacement thickness can be obtained by adding

$$\int_0^{\delta_B^*} \varrho_i u_i dy$$

to both sides of Eq. (5.1.5). We obtain

$$\int_0^{\delta_B^*} \varrho_i u_i dy = \varrho_{iw} u_{iw} \delta_A^* \tag{5.1.6}$$

Thus the transpiration condition given by Eq. (5.1.4) is equivalent to

$$\varrho_{iw} v_{iw} = \frac{d}{dx} \int_0^{\delta_B^*} \varrho_i u_i dy \tag{5.1.7}$$

which provides a physical explanation of it; that the mass flow injected normal to the wall in the equivalent inviscid flow, in order to displace the dividing streamline outwards the required distance δ_B^* from the surface, must be equal to the streamwise rate of change of the mass flow deficit in the boundary layer in the range $0 < y < \delta_B^*$.

In summary we see that there are two alternative inner boundary conditions which can be used in the calculation of the equivalent inviscid flow, namely, (1) a transpiration condition of nonzero normal velocity at the surface, Eq. (5.1.4), or (2) the condition that the displacement surface is a streamline, Eq. (5.1.5). Both mathematical models are equally valid and general representations of the displacement effect of the boundary layer on the external flow. In practice, however, the transpiration model is more convenient to use. For example, if a panel method is used to calculate the inviscid flow, the matrix of influence coefficients (Section 5.2) can be set up and inverted once and for all, so that subsequent inviscid calculations become trivial, while with a finite difference or finite volume method a fixed computational grid can be used throughout. For this reason, we will only use the transpiration model here.

Matching Conditions in the Wake

With the assumptions inherent in first-order boundary-layer theory, the conditions that are needed to effect a matching between the viscous and inviscid calculations to simulate a turbulent wake with the transpiration model are similar to those discussed for the conditions on the airfoil surface. A dividing streamline is chosen in the wake to separate the upper and lower parts of the inviscid flow, and on this line discontinuities are required in the normal components of velocity, so that it can be thought of as a source sheet.

At points C and D on the upper and lower sides of the dividing streamline (Fig. 5.3), the components of transpiration velocity, v_{iu} and v_{il} are, respectively, see Eq. (5.1.4),

$$v_{iu} = \frac{1}{\varrho_{iu}} \frac{d}{dx} (\varrho_{iu} u_{iu} \delta_u^*) \tag{5.1.8}$$

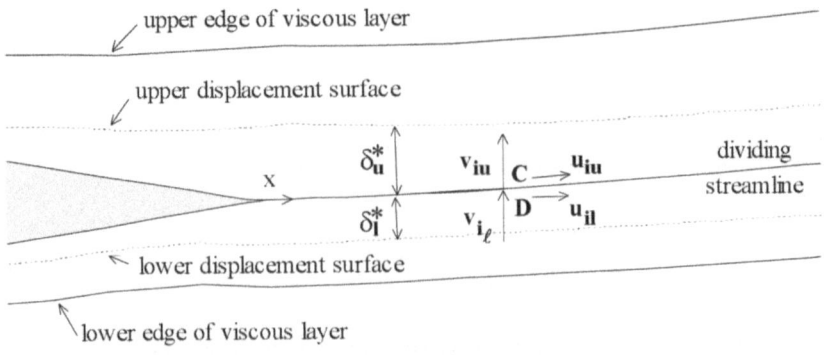

Fig. 5.3. Notation for the airfoil trailing-edge region.

and

$$v_{il} = -\frac{1}{\varrho_{il}} \frac{d}{dx}(\varrho_{il} u_{il} \delta_l^*)$$ (5.1.9)

Here the sign convention has been used that v is measured positive in the direction of the upward normal to the wake. Hence a jump Δv in the component of velocity normal to the wake is required; it is given by

$$\Delta v_i \equiv v_{iu} - v_{il} = \frac{1}{\varrho_{iu}} \frac{d}{dx}(\varrho_{iu} u_{iu} \delta_u^*) + \frac{1}{\varrho_{il}} \frac{d}{dx}(\varrho_{il} u_{il} \delta_l^*)$$ (5.1.10)

5.2 HS Panel Method

We consider an airfoil at rest in an onset flow of velocity V_∞. We assume that the airfoil is at an angle of attack, α (the angle between its chord line and the onset velocity), and that the upper and lower surfaces are given by functions $Y_u(x)$ and $Y_l(x)$, respectively. These functions can be defined analytically, or (as is often the case) by a set of (x, y) values of the airfoil coordinates. We denote the distance of any field point (x, y) from an arbitrary point, b, on the airfoil surface by r, as shown in Fig. 5.4. Let $\vec{n}$ also denote the unit vector normal to the airfoil surface and directed from the body into the fluid, and $\vec{t}$ a unit vector tangential to the surface, and assume that the inclination of $\vec{t}$ to the x-axis is given by θ. It follows from Fig. 5.4 that with $\vec{i}$ and $\vec{j}$ denoting unit vectors in the x- and y-directions, respectively,

$$\vec{n} = -\sin\theta\,\vec{i} + \cos\theta\,\vec{j}$$
$$\vec{t} = \cos\theta\,\vec{i} + \sin\theta\,\vec{j}$$ (5.2.1)

If the airfoil contour is divided into a large number of small segments, ds, then we can write

$$dx = \cos\theta\,ds$$
$$dy = \sin\theta\,ds$$ (5.2.2)

with ds calculated from $ds = \sqrt{(dx)^2 + (dy)^2}$.

We next assume that the airfoil geometry is represented by a finite number (N) of short straight-line elements called panels, defined by $(N+1)(x_j, y_j)$ pairs called boundary points. It is customary to input the (x, y) coordinates starting at the lower surface trailing edge, proceeding clockwise around the airfoil, and ending back at the upper surface trailing edge. If we denote the boundary points by

$$(x_1, y_1), (x_2, y_2), \ldots, (x_N, y_N), (x_{N+1}, y_{N+1})$$ (5.2.3)

then the pairs (x_1, y_1) and (x_{N+1}, y_{N+1}) are identical for a closed trailing edge (but not for an open trailing edge) and represent the trailing edge. It is customary to refer to the element between (x_j, y_j) and (x_{j+1}, y_{j+1}) as the j-th panel,

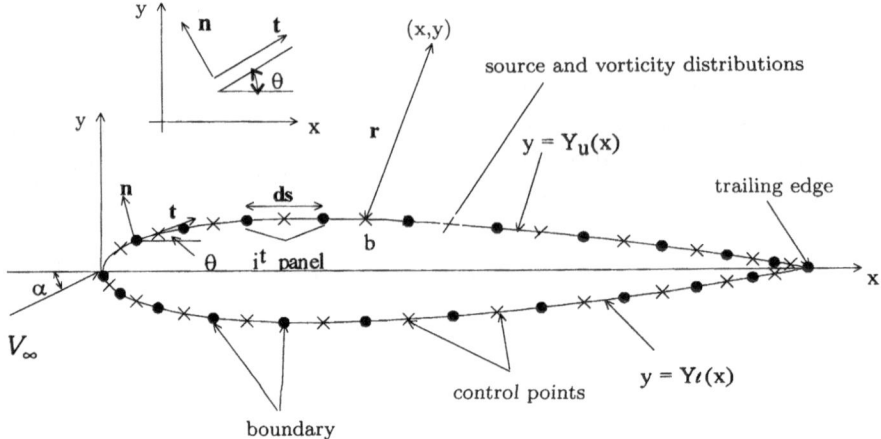

Fig. 5.4. Panel representation of airfoil surface and notation for an airfoil at incidence α.

and to the midpoints of the panels as the *control points*. Note from Fig. 5.4 that as one traverses from the i-th boundary point to the $(i+1)$-th boundary point, the airfoil body is on the right-hand side. This numbering sequence is consistent with the common definition of the unit normal vector $\vec{n}_i$ and unit tangential vector $\vec{t}_i$ for all panel surfaces, i.e., $\vec{n}_i$ is directed from the body into the fluid and $\vec{t}_i$ from the i-th boundary point to the $(i+1)$-th boundary point with its inclination to the x-axis given by θ_i.

In the HS panel method, the velocity $\vec{V}$ at any point (x, y) is represented by

$$\vec{V} = \vec{U} + \vec{v} \tag{5.2.4}$$

where $\vec{U}$ is the velocity of the uniform flow at infinity

$$\vec{U} = V_\infty (\cos\alpha\,\vec{i} + \sin\alpha\,\vec{j}) \tag{5.2.5}$$

and $\vec{v}$ is the disturbance field due to the body which is represented by two elementary flows corresponding to source and vortex flows. A source or vortex on the j-th panel causes an induced source velocity $\vec{v}_s$ at (x, y) or an induced vortex velocity $\vec{v}_v$ at (x, y), respectively, and these are obtained by taking gradients of a potential source

$$\phi_s = \frac{q}{2\pi} \ln r \tag{5.2.6}$$

and a potential vortex

$$\phi_r = -\frac{I'}{2\pi}\theta, \tag{5.2.7}$$

both centered at the origin, so that, with integrals applied to the airfoil surface,

$$\vec{v}(x, y) = \int \vec{v}_s q_j(s)ds_j + \int \vec{v}_v \tau_j(s)ds_j \tag{5.2.8}$$

Here $q_j ds_j$ is the source strength for the element ds_j on the j-th panel. Similarly, $\tau_j ds_j$ is the vorticity strength for the element ds_j on the same panel.

Each of the N panels is represented by similar sources and vortices distributed on the airfoil surface. The induced velocities in Eq. (5.2.8) satisfy the irrotationality condition and the boundary condition at infinity

$$u = \frac{\partial \phi}{\partial x} = \frac{\partial \psi}{\partial y} = V_\infty \cos \alpha \qquad (5.2.9a)$$

$$v = \frac{\partial \phi}{\partial y} = -\frac{\partial \psi}{\partial x} = V_\infty \sin \alpha \qquad (5.2.9b)$$

For uniqueness of the solutions, it is also necessary to specify the magnitude of the circulation around the body. To satisfy the boundary conditions on the body, which correspond to the requirement that the surface of the body is a streamline of the flow, that is,

$$\psi = \text{constant} \quad \text{or} \quad \frac{\partial \phi}{\partial n} = 0 \qquad (5.2.10)$$

at the surface on which n is the direction of the normal, the sum of the source-induced and vorticity-induced velocities and freestream velocity is set to zero in the direction normal to the surface of each of the N panels. It is customary to choose the *control points* to numerically satisfy the requirement that the resultant flow is tangent to the surface. If the tangential and normal components of the total velocity at the control point of the i-th panel are denoted by $(V^t)_i$ and $(V^n)_i$, respectively, the flow tangency conditions are then satisfied at panel control points by requiring that the resultant velocity at each control point has only $(V^t)_i$, and

$$(V^n)_i = 0 \quad i = 1, 2, \ldots, N \qquad (5.2.11)$$

Thus, to solve the Laplace equation with this approach, at the i-th panel control point we compute the normal $(V^n)_i$ and tangential $(V^t)_i$, $(i = 1, 2, \ldots, N)$ velocity components induced by the source and vorticity distributions on all panels, j ($j = 1, 2, \ldots, N$), including the i-th panel itself, and separately sum all the induced velocities for the normal and tangential components together with the freestream velocity components. The resulting expressions, which satisfy the irrotationality condition, must also satisfy the boundary conditions discussed above. Before discussing this aspect of the problem, it is convenient to write Eq. (5.2.4) expressed in terms of its velocity components $(V^n)_i$ and $(V^t)_i$ by

$$(V^n)_i = \sum_{j=1}^{N} A_{ij}^n q_j + \sum_{j=1}^{N} B_{ij}^n \tau_j + V_\infty \sin(\alpha - \theta_i) \qquad (5.2.12a)$$

$$(V^t)_i = \sum_{j=1}^{N} A_{ij}^t q_j + \sum_{j=1}^{N} B_{ij}^t \tau_j + V_\infty \cos(\alpha - \theta_i) \qquad (5.2.12b)$$

where A_{ij}^n, B_{ij}^n, A_{ij}^t, B_{ij}^t are known as influence coefficients, defined as the velocities induced at a control point (x_{m_i}, y_{m_i}); more specifically, A_{ij}^n and A_{ij}^t denote the normal and tangential velocity components, respectively, induced at the i-th panel control point by a unit strength source distribution on the j-th panel, and B_{ij}^n and B_{ij}^t are those induced by unit strength vorticity distribution on the j-th panel. The influence coefficients are related to the airfoil geometry and the panel arrangement; they are given by the following expressions:

$$A_{ij}^n = \begin{cases} \dfrac{1}{2\pi} \left[\sin(\theta_i - \theta_j) \ln \dfrac{r_{i,j+1}}{r_{i,j}} + \cos(\theta_i - \theta_j)\beta_{ij} \right] & i \neq j \\ \dfrac{1}{2} & i = j \end{cases} \tag{5.2.13}$$

$$A_{ij}^t = \begin{cases} \dfrac{1}{2\pi} \left[\sin(\theta_i - \theta_j)\beta_{ij} - \cos(\theta_i - \theta_j) \ln \dfrac{r_{i,j+1}}{r_{i,j}} \right] & i \neq j \\ 0 & i = j \end{cases} \tag{5.2.14}$$

$$B_{ij}^n = -A_{ij}^t \quad B_{ij}^t = A_{ij}^n \tag{5.2.15}$$

Here

$$r_{i,j+1} = \left[(x_{m_i} - x_{j+1})^2 + (y_{m_i} - y_{j+1})^2 \right]^{1/2}$$

$$r_{i,j} = \left[(x_{m_i} - x_j)^2 + (y_{m_i} - y_j)^2 \right]^{1/2}$$

$$x_{m_i} = \frac{1}{2}(x_i + x_{j+1}), \quad y_{m_i} = \frac{1}{2}(y_i + y_{j+1}) \tag{5.2.16}$$

$$\theta_i = \tan^{-1}\left(\frac{y_{i+1} - y_i}{x_{i+1} - x_i} \right), \quad \theta_j = \tan^{-1}\left(\frac{y_{j+1} - y_j}{x_{j+1} - x_j} \right)$$

$$\beta_{ij} = \tan^{-1}\left(\frac{y_{m_i} - y_{j+1}}{x_{m_i} - x_{j+1}} \right) - \tan^{-1}\left(\frac{y_{m_i} - y_j}{x_{m_i} - x_j} \right)$$

Regardless of the nature of $q_j(s)$ and $\tau_j(s)$, Eq. (5.2.12) satisfies the irrotationality condition and the boundary condition at infinity, Eq. (5.2.9). To satisfy the requirements given by Eq. (5.2.11) and the condition related to the circulation, it is necessary to adjust these functions. In the approach adopted by Hess and Smith [1], the source strength $q_j(s)$ is assumed to be constant over the j-th panel and is adjusted to give zero normal velocity over the airfoil, and the vorticity strength τ_j is taken to be constant on all panels ($\tau_j = \tau$) and its single value is adjusted to satisfy the condition associated with the specification of circulation. Since the specification of the circulation renders the solution unique, a rational way to determine the solution is required.

The best approach is to adjust the circulation to give the correct force on the body as determined by experiment. However, this requires advance knowledge of that force, and one of the principal aims of a flow calculation method is to calculate the force and not to take it as given. Thus, another criterion for determining circulation is needed.

For smooth bodies such as ellipses, the problem of rationally determining the circulation has yet to be solved. Such bodies have circulation associated with them, and resulting lift forces, but there is no rule for calculating these forces. If, on the other hand, we deal with an airfoil having a sharp trailing edge, we can apply the Kutta condition [5, 6]. It turns out that for every value of circulation except one, the inviscid velocity is infinite at the trailing edge. The Kutta condition states that the particular value of circulation that gives a finite velocity at the trailing edge is the proper one to choose. This condition does not include bodies with nonsharp trailing edges and bodies on which the viscous effects have been simulated by, for example, surface blowing, as discussed [4]. Thus, the classical Kutta condition is of strictly limited validity. It is customary to apply a "Kutta condition" to bodies outside its narrow definition, but this is an approximation; nevertheless the calculations are often in close accord with experiment.

In the panel method, the Kutta condition is indirectly applied by deducing another property of the flow at the trailing edge that is a direct consequence of the finiteness of velocity; this property is used as "the Kutta condition." Properties that have been used in lieu of "the Kutta condition" in panel methods include the following:

(a) A streamline of the flow leaves the trailing edge along the bisector of the trailing-edge angle.
(b) Upper and lower displacement total velocities approach a common limit at the trailing edge. The limiting value is zero if the trailing-edge angle is nonzero.
(c) Source and/or vorticity strengths at the trailing edge must satisfy conditions to allow finite velocity.

Of the above, property (b) is more widely used. At first it may be thought that this property requires setting both the upper and lower surface velocities equal to zero. This gives two conditions, which cannot be satisfied by adjusting a single parameter. The most reasonable choice is to make these two total velocities in the downstream direction at the 1st and N-th panel control points equal so that the flow leaves the trailing edge smoothly. Since the normal velocity on the surface is zero according to Eq. (5.2.11), the magnitudes of the two tangential velocities at the trailing edge must be equal to each other, that is,

$$(V^t)_N = -(V^t)_1 \tag{5.2.17}$$

Introducing the flow tangency condition, Eq. (5.2.11), into Eq. (5.2.12a) and noting that $\tau_j = \tau$, we get

$$\sum_{j=1}^{N} A_{ij}^n q_j + \tau \sum_{j=1}^{N} B_{ij}^n + V_\infty \sin(\alpha - \theta_i) = 0, \quad i = 1, 2, \ldots, N \tag{5.2.18}$$

In terms of the unknowns, q_j $(j = 1, 2, \ldots, N)$ and τ, the Kutta condition of Eq. (5.2.17) and Eq. (5.2.18) for a system of algebraic equations which can be written in the following form,

$$A\underset{\sim}{x} = \underset{\sim}{b} \tag{5.2.19}$$

Here A is a square matrix of order $(N + 1)$, that is

$$A \equiv \begin{vmatrix} a_{11} & a_{12} & \cdots & a_{1j} & \cdots & a_{1N} & a_{1,N+1} \\ a_{21} & a_{22} & \cdots & a_{2j} & \cdots & a_{2N} & a_{2,N+1} \\ \vdots & \vdots & \vdots & \vdots & \vdots & \vdots & \vdots \\ a_{i1} & a_{i2} & \cdots & a_{ij} & \cdots & a_{iN} & a_{i,N+1} \\ \vdots & \vdots & \vdots & \vdots & \vdots & \vdots & \vdots \\ a_{N1} & a_{N2} & \cdots & a_{Nj} & \cdots & a_{NN} & a_{N,N+1} \\ a_{N+1,1} & a_{N+1,2} & \cdots & a_{N+1,j} & \cdots & a_{N+1,N} & a_{N+1,N+1} \end{vmatrix} \tag{5.2.20}$$

and $\vec{x} = (q_1, \ldots, q_i, \ldots, q_N, \tau)^T$ and $b = (b_1, \ldots, b_i, \ldots, b_N, b_{N+1})^T$ with denoting the transpose. The elements of the coefficient matrix A follow from Eq. (5.2.18)

$$a_{ij} = A_{ij}^n, \quad \begin{array}{l} i = 1, 2, \ldots, N \\ j = 1, 2, \ldots, N \end{array} \tag{5.2.21a}$$

$$a_{i,N+1} = \sum_{j=1}^{N} B_{ij}^n, \quad i = 1, 2, \ldots, N \tag{5.2.21b}$$

A_{ij}^n are given by Eq. (5.2.13) and B_{ij}^n by Eq. (5.2.15). The relation in Eq. (5.2.20) follows from the definition of $\vec{x}$ where τ is essentially x_{N+1}.

To find $a_{N+1,j}$ $(J = 1, \ldots, N)$ and $a_{N+1,N+1}$ in the coefficient matrix A, we use the Kutta condition and apply Eq. (5.2.17) to Eq. (5.2.12b) and, with τ as a constant, we write the resulting expression as

$$\sum_{j=1}^{N} A_{1j}^t q_j + \tau \sum_{j=1}^{N} B_{1j}^t + V_\infty \cos(\alpha - \theta_1)$$

$$= - \left[\sum_{j=1}^{N} A_{Nj}^t q_j + \tau \sum_{j=1}^{N} B_{Nj}^t + V_\infty \cos(\alpha - \theta_N) \right]$$

or as

$$\sum_{j=1}^{N} (A_{1j}^t + A_{Nj}^t) q_j + \tau \sum_{j=1}^{N} (B_{1j}^t + B_{Nj}^t) \tag{5.2.22b}$$
$$= -V_\infty \cos(\alpha - \theta_1) - V_\infty \cos(\alpha - \theta_N)$$

so that,

$$a_{N+1,j} = A_{1j}^t + A_{Nj}^t, \quad j = 1, 2, \ldots, N \tag{5.2.23a}$$

$$a_{N+1,N+1} = \sum_{j=1}^{N}(B_{1j}^t + B_{Nj}^t) \tag{5.2.23b}$$

where now A_{1j}^t and A_{Nj}^t are computed from Eq. (5.2.14) and B_{1j}^t and B_{Nj}^t from Eq. (5.2.15).

The components of $\vec{b}$ again follow from Eqs. (5.2.18) and (5.2.21). From Eq. (5.2.18),

$$b_i = -V_\infty \sin(\alpha - \theta_i), \quad i = 1, \ldots, N \tag{5.2.24a}$$

and from Eq. (5.2.22),

$$b_{N+1} = -V_\infty \cos(\alpha - \theta_1) - V_\infty \cos(\alpha - \theta_N) \tag{5.2.24b}$$

With all the elements of a_{ij} determined from Eqs. (5.2.21) and (5.2.23) and the elements of $\vec{b}$ from Eq. (5.2.24), the solution of Eq. (5.2.19) can be obtained by the Gaussian elimination method [7]. The elements of $\vec{x}$ are given by

$$x_i = \frac{1}{a_{ii}^{(i-1)}} - \left[b_i^{(i-1)} - \sum_{j=i+1}^{N+1} a_{ij}^{(i-1)} x_j \right] \quad i = N+1, \ldots, 1 \tag{5.2.25}$$

where

$$a_{ij}^{(k)} = a_{ij}^{(k-1)} - \frac{a_{ik}^{(k-1)}}{a_{kk}^{(k-1)}} a_{kj}^{(k-1)}, \quad \begin{array}{l} k = 1, \ldots, N \\ j = k+1, \ldots, N+1 \\ i = k+1, \ldots, N+1 \\ a_{ij}^{(0)} = a_{ij} \end{array} \tag{5.2.26a}$$

$$b_i^{(k)} = b_i^{(k-1)} - \frac{a_{ik}^{(k-1)}}{a_{kk}^{(k-1)}} b_k^{(k-1)}, \quad \begin{array}{l} k = 1, \ldots, N \\ i = k+1, \ldots, N+1 \\ b_i^{(0)} = b_i \end{array} \tag{5.2.26b}$$

5.3 Viscous Effects

The viscous effects can be introduced into the panel method by (1) replacing the zero normal-velocity condition, Eq. (5.2.11), by a nonzero normal-velocity condition $V_{iw}(x)$ and by (2) satisfying the Kutta condition, Eq. (5.2.17), not on the surface of the airfoil trailing edge but at some distance away from the surface.

Here it will be assumed that the nonzero normal-velocity distribution $V_{iw}(x)$ along the surface of the airfoil and in its wake is known, together with the distance from the surface, say displacement thickness δ^*, where the Kutta condition is to be satisfied. We now describe how these two new conditions can be incorporated into the panel method.

To include the nonzero normal-velocity condition into the solution procedure, we write Eq. (5.2.18) as

$$\sum_{j=1}^{N} A_{ij}^{n} q_j + \tau \sum_{j=1}^{N} B_{ij}^{n} q_j + V_\infty \sin(\alpha - \theta_i) = v_{iw}(x_{m_i}) \qquad (5.3.1)$$

To satisfy the Kutta condition at the normal distance δ^* from the surface of the trailing edge, called the "off-body" Kutta condition, the total velocities at the N-th and first off-body control points are again required to be equal. Since the normal velocity component is not zero, we write the off-body Kutta condition at distance δ^* as

$$(V)_N = -(V)_1 \qquad (5.3.2)$$

where V is the total velocity at the two control points. The off-body total velocities are computed from

$$V = \frac{(V^n)^2 + (V^t)^2}{V} = V^n \frac{V^n}{V} + V^t \frac{V^t}{V} = V^n \sin \phi + V^t \cos \phi \qquad (5.3.3)$$

where V^n and V^t are computed by expressions identical to those given by Eqs. (5.2.12) at the two off-body control points, $I = 1$, $I = N$, that is,

$$(V^n)_I = \sum_{j=1}^{N} A_{Ij}^{n} q_j + \tau \sum_{j=1}^{N} B_{Ij}^{n} + V_\infty \sin(\alpha - \theta_I) \qquad (5.3.4a)$$

$$(V^t)_I = \sum_{j=1}^{N} A_{Ij}^{t} q_j + \tau \sum_{j=1}^{N} B_{Ij}^{t} + V_\infty \cos(\alpha - \theta_I) \qquad (5.3.4b)$$

and where

$$\phi = \tan^{-1}[(V^n)_I / (V^t)_I] \qquad (5.3.5)$$

With Eqs. (5.3.4), the expression for the total velocity given by Eq. (5.3.3) can be written as

$$\begin{aligned} V = & \sum_{j=1}^{N} (A_{Ij}^{n} \cdot \sin \phi + A_{Ij}^{t} \cdot \cos \phi) q_j \\ & + \tau \sum_{j=1}^{N} (B_{Ij}^{n} \cdot \sin \phi + B_{Ij}^{t} \cdot \cos \phi) \\ & + V_\infty \sin(\alpha - \theta_I) \sin \phi + V_\infty \cos(\alpha - \theta_I) \cos \phi \end{aligned} \qquad (5.3.6a)$$

or as

$$V = \sum_{j=1}^{N} A_{Ij}' q_j + \tau \sum_{j=1}^{N} B_{Ij}' + V_\infty \cos(\alpha - \theta_I - \phi) \qquad (5.3.6b)$$

where

$$A'_{Ij} = A^n_{Ij} \cdot \sin\phi + A^t_{Ij} \cdot \cos\phi, \quad B'_{Ij} = B^n_{Ij} \cdot \sin\phi + B^t_{Ij} \cdot \cos\phi \tag{5.3.7a}$$

$$A^n_{Ij} = \frac{1}{2\pi}\left[\sin(\theta_I - \theta_j)\ln\frac{r_{I,j+1}}{r_{I,j}} + \cos(\theta_I - \theta_j)\beta_{Ij}\right] \tag{5.3.7b}$$

$$A^t_{Ij} = \frac{1}{2\pi}\left[\sin(\theta_I - \theta_j)\beta_{Ij} - \cos(\theta_I - \theta_j)\ln\frac{r_{I,j+1}}{r_{I,j}}\right] \tag{5.3.7c}$$

$$B^n_{Ij} = -A^t_{Ij}, \quad B^t_{Ij} = A^n_{Ij} \tag{5.3.7d}$$

If we define

$$\theta'_I = \theta_I + \phi \tag{5.3.8}$$

then it can be shown that Eq. (5.3.6b) can be written as

$$V = \sum_{j=1}^{N} A'_{Ij}q_j + \tau\sum_{j=1}^{N} B'_{Ij} + V_\infty\cos(\alpha - \theta'_I) \tag{5.3.9}$$

where

$$A'_{Ij} = \frac{1}{2\pi}\left[\sin(\theta'_I - \theta_j)\beta_{Ij} - \cos(\theta'_I - \theta_j)\ln\frac{r_{I,j+1}}{r_{I,j}}\right] \tag{5.3.10a}$$

$$B'_{Ij} = \frac{1}{2\pi}\left[\sin(\theta'_I - \theta_j)\ln\frac{r_{I,j+1}}{r_{I,j}} + \cos(\theta'_I - \theta_j)\beta_{Ij}\right] \tag{5.3.10b}$$

The off-body Kutta condition can now be expressed in a form similar to that of Eq. (5.2.22). Applying Eq. (5.3.2) to Eq. (5.3.9), we write the resulting expression as

$$\sum_{j=1}^{N} A'_{Nj}q_j + \tau\sum_{j=1}^{N} B'_{Nj} + V_\infty\cos(\alpha - \theta'_N)$$
$$= -\left[\sum_{j=1}^{N} A'_{1j}q_j + \tau\sum_{j=1}^{N} B'_{1j} + V_\infty\cos(\alpha - \theta'_1)\right] \tag{5.3.11a}$$

or as

$$\sum_{j=1}^{N}(A'_{1j} + A'_{Nj})q_j + \tau\sum_{j=1}^{N}(B'_{1j} + B'_{Nj})$$
$$+ V_\infty\cos(\alpha - \theta'_1) + V_\infty\cos(\alpha - \theta'_N) = 0 \tag{5.3.11b}$$

5.4 Flowfield Calculation in the Wake

As discussed in Section 4.2, the calculation of airfoils in incompressible viscous flows can be accomplished without taking into account the wake effect; that is, the viscous flow calculations are performed up to the trailing edge only and are not extended into the wake. This procedure, which may be sufficient at low to moderate angles of attack without flow separation, is not sufficient at higher angles of attack, including post-stall flows. Additional changes are required in the panel method (and in the boundary-layer method), as discussed in this section.

The viscous wake calculations usually include a streamline issuing from the trailing edge of the airfoil. The computation of the location of this streamline is relatively simple if conformal mapping methods are used to determine the velocity field. In this case, the stream function ψ is usually known, and because the airfoil surface is represented by $\psi(x, y) = \text{const}$, the calculation of the wake streamline amounts to tracing the curve after it leaves the airfoil. When the flowfield is computed by a panel method or by a finite-difference method, however, the results are known only at discrete points in the field in terms of the velocity components. In this case, the wake streamline is determined from the numerical integration of

$$\frac{dy}{dx} = \frac{v}{u} \tag{5.4.1}$$

aft of the trailing edge with known initial conditions. However, some care is necessary in selecting the initial conditions, especially when the trailing edge is blunt. As a general rule, the initial direction of the streamline is given to a good approximation by the bisector of the trailing-edge angle of the airfoil.

The panel method, which was modified only for an airfoil flow, now requires similar modifications to include the viscous effects in the wake which behaves as a distribution of sinks. It is divided into nwp panels along the dividing stream-line with suction velocities or sink strengths $q_i = \Delta v_i$ ($N + 1 \leq i \leq N + nwp$), distributed on the wake panels and determined from boundary-layer solutions in the wake by Eq. (5.2.12). As before, off-body boundary points and "control" points are introduced at the intersections of the δ^* surface with the normals through panel boundary points and panel control points, respectively. Summation of all the induced velocities, separately for the normal and tangential components and together with the freestream velocity components, produces $(V^n)_I$ and $(V^t)_I$ at $I = 1, 2, \ldots, N + nwp$. The wake velocity distribution, as the airfoil velocity distribution, is computed on the δ^*-surface, rather than on the dividing streamline.

The total velocities are again computed from Eq. (5.3.3), with $(V^n)_I$ and $(V^t)_I$ from Eqs. (5.3.4), except that now

$$(V^n)_I = \sum_{j=1}^{N+nwp} A^n_{Ij} q_j + \tau \sum_{j=1}^{N} B^n_{Ij} + V_\infty \sin(\alpha - \theta_I) \tag{5.4.2a}$$

$$(V^t)_I = \sum_{j=1}^{N+nwp} A^t_{Ij} q_j + \tau \sum_{j=1}^{N} B^t_{Ij} + V_\infty \cos(\alpha - \theta_I) \tag{5.4.2b}$$

As before, the expression for the total velocity is written in the same form as Eq. (5.3.6a), except that now

$$\begin{aligned} V = & \sum_{j=1}^{N+nwp} (A^n_{Ij} \cdot \sin\phi + A^t_{Ij} \cdot \cos\phi) q_j \\ & + \tau \sum_{j=1}^{N} (B^n_{Ij} \cdot \sin\phi + B^t_{Ij} \cdot \cos\phi) \\ & + V_\infty \sin(\alpha - \theta_I) \sin\phi + V_\infty \cos(\alpha - \theta_I) \cos\phi \end{aligned} \tag{5.4.3}$$

where A^n_{Ij}, A^t_{Ij}, B^n_{Ij} and B^t_{Ij}, are identical to those given by Eq. (5.3.7). Similarly, Eq. (5.3.9) with A_{Ij} and B_{Ij} given by Eq. (5.3.10) is

$$V = \sum_{j=1}^{N+nwp} A'_{Ij} q_j + \tau \sum_{j=1}^{N} B'_{Ij} + V_\infty \cos(\alpha - \theta'_I) \tag{5.4.4}$$

and the Kutta condition given by Eqs. (5.3.11a) becomes

$$\begin{aligned} & \sum_{j=1}^{N+nwp} A'_{Nj} q_j + \tau \sum_{j=1}^{N} B'_{Nj} + V_\infty \cos(\alpha - \theta'_N) \\ & = - \left[\sum_{j=1}^{N+nwp} A'_{1j} q_j + \tau \sum_{j=1}^{N} B'_{1j} + V_\infty \cos(\alpha - \theta'_1) \right] \end{aligned} \tag{5.4.5a}$$

or

$$\begin{aligned} & \sum_{j=1}^{N+nwp} (A'_{1j} + A'_{Nj}) q_j + \tau \sum_{j=1}^{N} (B'_{1j} + B'_{Nj}) \\ & + V_\infty \cos(\alpha - \theta'_1) + V_\infty \cos(\alpha - \theta'_N) = 0 \end{aligned} \tag{5.4.5b}$$

In computing the wake velocity distribution at distances δ^* from the wake dividing streamline, the velocities in the upper wake are equal to those in the lower wake for a symmetrical airfoil at zero angle of attack. This is, however, not the case if the airfoil is asymmetric or if the airfoil is at an angle of incidence. While the external velocities on the upper and lower surfaces at the trailing edge are equal to each other, they are not equal to each other in the wake region since the δ^*-distribution in the upper wake is different from the δ^*-distribution in the lower wake.

5.5 Computer Program

In this section we present a computer program for the panel method discussed in the previous sections. This program can be used interactively with the boundary-layer program of Chapter 4 so that, as discussed in detail in [4] and briefly in Section 5.6, more accurate solutions of inviscid and viscous flow equations can be obtained by including the viscous effects in the panel method of Section 5.2.

The computer program of the panel method has five subroutines and MAIN, as described below.

5.5.1 MAIN

MAIN contains the input information which comprises (1) the number of panels along the surface of the airfoil, NODTOT, and the number of panels in the wake, NW. The code is arranged so that it can be used for inviscid flows with and without viscous effects. For inviscid flows, NW is equal to zero. (2) The next input data also comprises airfoil coordinates normalized with respect to its chord c, x/c, y/c, $[\equiv X(I), Y(I)]$. If NW $\neq 0$, then it is necessary to specify the dimensionless displacement thickness δ^*/c ($\equiv$ DLSP(I)), dimensionless blowing velocity u_w/u_∞ ($\equiv$ VNP(I)) distributions on the airfoil, as well as the wake coordinates XW(I), YW(I) of the dividing streamline, the dimensionless displacement thickness distribution on the upper wake DELW(I,1) and lower wake DELW(I,2) and velocity jump QW(I). It should be noted that all input data for wake includes values at the trailing edge. The input also includes angle of attack α ($\equiv$ ALPHA) and Mach number M_∞ ($\equiv$ FMACH).

The panel slopes are calculated from Eq. (5.2.2). The subroutine COEF is called to compute A and $\vec{b}$ in Eq. (5.2.19) subroutine OBKUTA to calculate the off-body Kutta condition, subroutine GAUSS to compute $\vec{x}$, subroutine VPDIS to compute the velocity and pressure distributions, and subroutine CLCM to compute the airfoil characteristics corresponding to lift (CL) and pitching moment (CM) coefficients.

5.5.2 Subroutine COEF

This subroutine calculates the elements a_{ij} of the coefficient matrix A from Eqs. (5.2.21) and (5.2.23) and the elements of $\vec{b}$ from Eq. (5.2.24). We note that $N+1$ corresponds to KUTTA, and N to NODTOT

5.5.3 Subroutine OBKUTA

This subroutine is used to calculate the body-off Kutta condition.

5.5.4 Subroutine GAUSS

The solution of Eq. (5.2.19) is obtained with the Gauss elimination method described in Section 5.2.

5.5.5 Subroutine VPDIS

Once $\vec{x}$ is determined by subroutine GAUSS so that source strengths q_i ($i = 1, 2, \ldots, N$) and vorticity τ on the airfoil surface are known, the tangential velocity component (V^t) at each control point can be calculated. Denoting q_i with Q(I) and τ with GAMMA, the tangential velocities $(V^t)_i$ are obtained with the help of Eq. (5.2.12b). This subroutine also determines the distributions of the dimensionless pressure coefficient C_p ($\equiv CP$) defined by

$$C_p = \frac{p - p_\infty}{(1/2)\varrho V_\infty^2} \tag{5.5.1a}$$

which in terms of velocities can be written as

$$C_p = 1 - \left(\frac{V^t}{V_\infty}\right)^2 \tag{5.5.1b}$$

It is common to use panel methods for low Mach number flows by introducing compressibility corrections which depend upon the linearized form of the compressibility velocity potential equation and are based on the assumption of small perturbations and thin airfoils [5]. A simple correction formula for this purpose is the Karman-Tsien formula which uses the "tangent gas" approximation to simplify the compressible potential-flow equations. According to this formula, the effect of Mach number on the pressure coefficient is estimated from

$$c_p = \frac{c_{p_i}}{\beta + [M_\infty^2/(1 + \beta)](c_{p_i}/2)} \tag{5.5.2}$$

and the corresponding velocities are computed from

$$V^2 = 1 + \frac{1}{c_6}[1 - (1 + c_8 c_p)^{1/c_7}] \tag{5.5.3}$$

Here c_{p_i} denotes the incompressible pressure coefficient, M_∞ the freestream Mach number and

$$\beta = \sqrt{1 - M_\infty^2}, \quad c_6 = \frac{\gamma - 1}{2} M_\infty^2, \quad c_7 = \frac{\gamma}{\gamma - 1}, \quad c_8 = \frac{1}{2}\gamma M_\infty^2, \quad \gamma = 1.4$$

$$\tag{5.5.4}$$

In this subroutine we also include this capability in the HS panel method.

5.5.6 Subroutine CLCM

The dimensionless pressure in the appropriate directions is integrated to compute the aerodynamic force and the coefficients for lift (CL) and pitching moment (CM) about the leading edge of the airfoil.

5.5.7 Subroutine VPDWK

This subroutine calculates the total velocity and pressure coefficient at each control point along the upper and lower wakes separately. The normal and tangential components of the total velocities are computed from Eqs. (5.4.2a) and (5.4.2b).

References

[1]   Hess, J. L. and Smith, A. M. O., "Calculation of Potential Flow About Arbitrary Bodies," *Progress in Aerospace Sciences*, Vol. 8, Pergamon Press, N.Y., 1966.

[2]   Lighthill, M. J., "On Displacement Thickness," J. Fluid Mech., Vol. 4, p. 383, 1958.

[3]   Lock, R. C. and Firmin, M. C. P., "Survey of Techniques for Estimating Viscous Effects in External Aerodynamics," Royal Aircraft Establishment Tech Memo, AERO 1900, 1981.

[4]   Cebeci, T., *An Engineering Approach to the Calculation of Aerodynamic Flows*, Horizons Publishing, Long Beach, CA and Springer, Heidelberg, Germany, 1999.

[5]   Anderson, J., *Aerodynamics*, McGraw-Hill, NY, 1988.

[6]   Moran, J., *An Introduction to Theoretical and Computational Aerodynamics*, John Wiley, NY, 1984.

[7]   Isaacson, E. and Keller, H. B., *Analysis of Numerical Methods*, John Wiley, NY, 1966.

6 Application of the Computer Program for the CS and k-ε Models to Other Higher-Order Turbulence Models

6.0 Introduction

The computer program for CS and k-ε models described in Chapter 3 can also be used to obtain the solution of the boundary-layer equations with other turbulence models, including flows with separation. Here we first consider non-separating flows; then we discuss the application of the computer program of Chapter 3 to k-ω and SST models in Section 6.1 and to the SA model in Section 6.2. In Section 6.3 we present a brief description of the extension of the k-ε model to flows with separation.

6.1 Solution of the k-ω and SST Model Equations

The solution of the k-ω model equations with the computer program of Chapter 3 is similar to the solution of the k-ε model equations with wall functions. Again the k-ω model equations, Eqs. (1.2.18), (1.3.10), and (1.3.11), are expressed in terms of Falkner-Skan variables.

Since the SST model equations make use of the k-ω model equations in the inner region and the k-ε model equations in the outer region we express them, for the sake of compactness, in the following form in transformed variables

$$[(1+\sigma_k\varepsilon_m^+)k']' - 2mf'k + m_1 fk' + \varepsilon_m^+(f'')^2 - \beta^* \omega k = x\left(f'\frac{\partial k}{\partial x} - k'\frac{\partial f}{\partial x}\right) \quad (6.1.1)$$

$$[(1+\sigma_\omega\varepsilon_m^+)\omega']' + 2(1-F_1)\sigma_{\omega_2}\frac{R_x}{\omega}k'\omega' + m_1\omega'f - (m-1)f'\omega$$
$$- \beta\omega^2 + R_x(f'')^2 = x\left(f'\frac{\partial \omega}{\partial x} - \omega'\frac{\partial f}{\partial x}\right) \quad (6.1.2)$$

where ω and k are dimensionless, normalized by x/u_e and $1/u_e^2$, respectively. Equations (6.1.1) and (6.1.2) are the equations used in the SST model. To recover Wilcox's k-ω model equations expressed in transformed variables, we let $F_1 = 1$ and take

$$\sigma_k = 0.5, \quad \sigma_\omega = 0.5, \quad \beta = 0.075,$$

$$\beta^* = 0.09, \quad \kappa = 0.41, \quad \gamma = \frac{\beta}{\beta^*} - \sigma_\omega \frac{\kappa^2}{\sqrt{\beta^*}} \tag{6.1.3}$$

In the SST model, the above constants are determined from the relation, Eq. (1.4.10),

$$\phi = F_1\phi_1 + (1 - F_1)\phi_2 \tag{1.4.10}$$

where the constant ϕ_1 is determined from Eq. (1.4.11) and the constant ϕ_2 from Eq. (1.4.12). F_1 is detemined from Eq. (1.4.6), where its $\arg_1$ given by Eq. (1.4.7) can be written as

$$\arg_1 = \min[\max(\lambda_1, \lambda_2), \lambda_3] \tag{6.1.4}$$

In terms of transformed quantities, λ_1 to λ_3 and $CD_{k\omega}$ are

$$\lambda_1 = \frac{\sqrt{k}}{0.09\omega y} = \frac{\sqrt{k}\,\sqrt{R_x}}{\omega\eta\,0.09} \tag{6.1.5a}$$

$$\lambda_2 = \frac{500\nu}{y^2\omega} = \frac{500}{\eta^2\omega} \tag{6.1.5b}$$

$$\lambda_3 = \frac{4\varrho\sigma_{\omega 2}k}{CD_{k\omega}y^2} \tag{6.1.5c}$$

$$CD_{k\omega} = \max\left(2\varrho\sigma_{\omega 2}\frac{1}{\omega}\frac{\partial k}{\partial y}\frac{\partial\omega}{\partial y}, 10^{-20}\right) = \frac{2k}{\max(\frac{1}{\omega}k'\omega', 10^{-20})\eta^2}$$

We first find the maximum of λ_1 and λ_2 (say λ_4), then calculate the minimum of λ_4 and λ_3 and thus determine $\arg_1$ and F_1. Once F_1 is calculated, then the constants in Eqs. (6.1.1) and (6.1.2) are determined from the relation given by Eq. (1.4.10). For example,

$$\sigma_\omega = 0.5F_1 + 0.856(1 - F_1)$$

$$\beta = 0.0750F_1 + 0.0828(1 - F_1) \text{ etc.}$$

Next we determine the eddy viscosity distribution across the boundary layer. In terms of transformed variables, Eq. (1.4.1) can be written as ($v = f''$, $a_1 = 0.31$)

$$\varepsilon_m^+ \equiv \frac{\varepsilon_m}{\nu} = \begin{cases} R_x\dfrac{k}{\omega} & a_1\omega > \Omega F_2 \\[2ex] \dfrac{\sqrt{R_x}\,a_1 k}{|v|F_2} & a_1\omega < \Omega F_2 \end{cases} \tag{6.1.6}$$

where

$$\Omega = \left|\frac{\partial u}{\partial y}\right| = u_e|f''|\sqrt{\frac{u_e}{\nu x}}$$

and F_2 is determined from Eq. (1.4.2a) where arg_2 is

$$\mathrm{arg}_2 = \max(2\lambda_1, \lambda_2) \tag{6.1.7}$$

In the SST model, once the constants are determined and the distribution of eddy viscosity is calculated, then Eqs. (6.1.1) and (6.1.2) are solved together with the continuity and momentum equations; a new arg_1, arg_2, F_1 and F_2, new constants and eddy viscosity distribution are determined. This procedure is repeated until convergence.

It should be noted that, for $F_1 = 1$, the whole region is the inner region governed by the k-ω model equations. When $F_1 = 0$, the whole region is governed by the k-ε model equations.

Before we discuss the solution procedure for the SST model equations, it is useful to point out that the structure of the solution algorithm for the k-ε model equations with wall functions is almost identical to the one for the SST model equations. This means all the A_j, B_j, C_j matrices have the same structure; the difference occurs in the definitions of the coefficients of the linearized momentum, kinetic energy and rate-of-dissipation equations and in the definition of the boundary condition for ω which occurs in the fourth row of A_0-matrix.

To describe the numerical method for the k-ω model equations, we start with the kinetic energy equation, Eq. (6.1.1), and write it in the same form as Eq. (2.1.4) by defining Q and F by

$$Q = \beta^* \omega k, \qquad F = 0 \tag{6.1.8}$$

The definition of P remains the same. Next we write Eq. (6.1.2) in the form

$$(b_3\omega')' + P_1 - Q_1 + E = x\left(f'\frac{\partial \omega}{\partial x} - \omega'\frac{\partial f}{\partial x}\right) + (m-1)f'\omega - m_1\omega'f \tag{6.1.9}$$

where

$$E = 2(1 - F_1)\sigma_{\omega 2}\frac{R_x}{\omega}k'\omega'$$

$$Q_1 = \beta\omega^2 \tag{6.1.10}$$

$$P_1 = \gamma R_x(f'')^2$$

With

$$\omega' = q$$

Equation (6.1.9) can be written as

$$(b_3 q)' + P_1 - Q_1 + E = x\left(u\frac{\partial \omega}{\partial x} - q\frac{\partial f}{\partial x}\right) + (m-1)u\omega - m_1 qf \tag{6.1.11}$$

A comparison of Eq. (6.1.11) with Eq. (2.2.4) shows that if we let $\varepsilon = \omega$ for notation purposes, then the coefficients of linearized specific dissipation rate equation are very similar to those given by Eqs. (3.3.8) and (3.3.9). Except for the definitions of Q and F in the kinetic energy-equation, Eq. (6.1.1), the coefficients of the linearized kinetic energy equation are identical to those given by Eqs. (3.3.4) and (3.3.5). Appropriate changes then can be easily made to subroutine KECOEF in order to adopt the computer program of Chapter 3 to solve the kinetic energy and specific dissipation rate kinetic energy equations in the SST model. Of course, other changes also should be made, but these are not discussed here. A good understanding of the computer program for the k-ε model equations is needed to make the necessary changes.

6.2 Solution of the SA Model Equations

The solution of the SA model equations with the computer program of Chapter 3 is similar to the solution of the k-ε and SST model equations. As before we again express the equations in transformed variables. Since this model involves the solution of the continuity and momentum equations with the transport equation for eddy viscosity, Eq. (1.5.4), and since the continuity and momentum equations are already transformed we only need to transform Eq. (1.5.4). In terms of the Falkner-Skan variables, this equation can be written as

$$
x\left(f'\frac{\partial \nu_t^+}{\partial x} - (\nu_t^+)'\frac{\partial f}{\partial x} \right) - m_1(\nu_t^+)'f = c_{b_1}(1 - f_{t_2})\tilde{s}^* \nu_t^+
$$

$$
- \left(c_{w1}f_w - \frac{c_{b_1}}{\kappa^2}f_{t_2} \right)\frac{(\nu_t^+)^2}{\eta^2 + (\nu_w^*)^2}
$$

$$
+ \frac{1}{\sigma}[[(1 + \nu_t^+)(\nu_t^+)']' + c_{b_2}[(\nu_t^+)']^2] \tag{6.2.1}
$$

We now define

$$
g = (\nu_t^+)', \quad u = f', \quad v = u' \tag{6.2.2}
$$

and write the transformed equations (2.1.3) and (6.2.1) as

$$
f' = u \tag{6.2.3a}
$$

$$
u' = v \tag{6.2.3b}
$$

$$
(\nu_t^+)' = g \tag{6.2.3c}
$$

$$
(bv)' + m_1 fv + m(1 - u^2) = x\left(u\frac{\partial u}{\partial x} - v\frac{\partial f}{\partial x} \right) \tag{6.2.3d}
$$

$$\frac{1}{\sigma}[[(1+\nu_t^+)g]' + c_{b_2}g^2] + c_{b_1}(1-f_{t_2})\tilde{s}^*\nu_t^+$$

$$-\left(c_{w_1}f_w - \frac{c_{b_1}}{\kappa^2}f_{t_2}\right)\frac{(\nu_t^+)^2}{\eta^2 + (\nu_w^*)^2} + m_1gf =$$

$$x\left(u\frac{\partial\nu_t^+}{\partial x} - g\frac{\partial f}{\partial x}\right)$$

(6.2.3e)

Using the numerical method discussed in Chapter 2, the above equations can be expressed in the following linearized form (note $\nu_t^+ \equiv \nu^+$ for convenience)

$$\delta f_j - \delta f_{j-1} - \frac{h_j}{2}(\delta u_j + \delta u_{j-1}) = (r_1)_j$$

(6.2.4a)

$$\delta u_j - \delta u_{j-1} - \frac{h_j}{2}(\delta v_j + \delta v_{j-1}) = (r_4)_{j-1}$$

(6.2.4b)

$$\delta\nu_j^+ - \delta\nu_{j-1}^+ - \frac{h_j}{2}(\delta g_j + \delta g_{j-1}) = (r_5)_{j-1}$$

(6.2.4c)

$$(s_1)_j\delta f_j + (s_2)_j\delta f_{j-1} + (s_3)_j\delta u_j + (s_4)_j\delta u_{j-1} + (s_5)_j\delta v_j$$
$$+ (s_6)_j\delta v_{j-1} + (s_7)_j\delta\nu_j^+ + (s_8)_j\delta\nu_{j-1}^+ = (r_2)_j$$

(6.2.4d)

$$(e_1)_j\delta f_j + (e_2)_j\delta f_{j-1} + (e_3)_j\delta u_j + (e_4)_j\delta u_{j-1} + (e_5)_j\delta v_j$$
$$+ (e_6)_j\delta v_{j-1} + (e_7)_j\delta\nu_j^+ + (e_8)_j\delta\nu_{j-1}^+ + (e_9)_j\delta g_j$$
$$+ (e_{10})_j\delta g_{j-1} = (r_3)_j$$

(6.2.4e)

Here the coefficients of Eq. (6.2.4d) are

$$(s_1)_j = \tilde{\alpha}v_{j-1/2} + \frac{m_1}{2}v_j$$

(6.2.5a)

$$(s_2)_j = \tilde{\alpha}v_{j-1/2} + \frac{m_1}{2}v_{j-1}$$

(6.2.5b)

$$(s_3)_j = -(m+\tilde{\alpha})u_j$$

(6.2.5c)

$$(s_4)_j = -(m+\tilde{\alpha})u_{j-1}$$

(6.2.5d)

$$(s_5)_j = b_j h_j^{-1} + \frac{m_1}{2}f_j + 0.5x\left(\frac{\partial f}{\partial x}\right)_{j-1/2}$$

(6.2.5e)

$$(s_6)_j = -b_{j-1}h_j^{-1} + \frac{m_1}{2}f_{j-1} + 0.5x\left(\frac{\partial f}{\partial x}\right)_{j-1/2}$$

(6.2.5f)

$$(s_7)_j = \left(\frac{\partial b}{\partial\nu^+}\right)_j v_j h_j^{-1}$$

(6.2.5g)

$$(s_8)_j = -\left(\frac{\partial b}{\partial\nu^+}\right)_{j-1} v_{j-1}h_j^{-1}$$

(6.2.5h)

$$(r_2)_j = x \left[\frac{1}{2} \left(\frac{\partial u^2}{\partial x} \right)_{j-1/2} - v_{j-1/2} \left(\frac{\partial f}{\partial x} \right)_{j-1/2} \right]$$
$$- \left[\frac{b_j v_j - b_{j-1} v_{j-1}}{h_j} + m_1(fv)_{j-1/2} + m(1 - u_{j-1/2}^2) \right] \quad (6.2.6)$$

where with A_1, A_2, A_3 given by Eq. (2.3.9),

$$m_1 = \frac{m+1}{2}, \quad \tilde{\alpha} = \frac{1}{2} A_3 x \quad (6.2.7a)$$

$$\left(\frac{\partial f}{\partial x} \right)_{j-1/2}^n = A_1 f_{j-1/2}^{n-2} + A_2 f_{j-1/2}^{n-1} + A_3 f_{j-1/2}^n \quad (6.2.7b)$$

The term $\frac{\partial b}{\partial \nu^+}$ denotes the variation of b with respect to ν^+ and is given by

$$\frac{\partial b}{\partial \nu^+} = f_{v_1} + \nu^+ \frac{\partial f_{v_1}}{\partial \nu^+} \quad (6.2.8)$$

The coefficients of Eq. (6.2.4e) are

$$e_1 = \tilde{\alpha} g_{j-1/2} + \frac{m_1}{2} g_j$$

$$e_2 = \tilde{\alpha} g_{j-1/2} + \frac{m_1}{2} g_{j-1}$$

$$e_3 = -\frac{1}{2} x \left(\frac{\partial \nu^+}{\partial x} \right)_{j-1/2}, \quad e_4 = e_3$$

$$e_5 = \frac{1}{2} \left[\left(\frac{\partial \tilde{p}_r}{\partial v} \right)_j - \left(\frac{\partial \tilde{d}_e}{\partial v} \right)_j \right]$$

$$e_6 = \frac{1}{2} \left[\left(\frac{\partial \tilde{p}_r}{\partial v} \right)_{j-1} - \left(\frac{\partial \tilde{d}_e}{\partial v} \right)_{j-1} \right]$$

$$e_7 = -\tilde{\alpha} u_{j-1/2} + \left(\frac{\partial \tilde{f}_\mu}{\partial \nu^+} \right)_j + \frac{1}{2} \left[\left(\frac{\partial \tilde{p}_r}{\partial \nu^+} \right)_j - \left(\frac{\partial \tilde{d}_e}{\partial \nu^+} \right)_j \right]$$

$$e_8 = -\tilde{\alpha} u_{j-1/2} + \left(\frac{\partial \tilde{f}_\mu}{\partial \nu^+} \right)_{j-1} + \frac{1}{2} \left[\left(\frac{\partial \tilde{p}_r}{\partial \nu^+} \right)_{j-1} - \left(\frac{\partial \tilde{d}_e}{\partial \nu^+} \right)_{j-1} \right]$$

$$e_9 = \frac{1}{2} x \left(\frac{\partial f}{\partial x} \right)_{j-1/2} + \frac{m_1}{2} f_j + \left(\frac{\partial \tilde{f}_\mu}{\partial g} \right)_j$$

$$e_{10} = \frac{1}{2} x \left(\frac{\partial f}{\partial x} \right)_{j-1/2} + \frac{m_1}{2} f_{j-1} - \left(\frac{\partial \tilde{f}_\mu}{\partial g} \right)_{j-1}$$

$$(r_3)_j = \left[u_{j-1/2}x \left(\frac{\partial \nu^+}{\partial x} \right)_{j-1/2} - g_{j-1/2}x \left(\frac{\partial f}{\partial x} \right)_{j-1/2} - m_1 (fg)_{j-1/2} \right]$$
$$- (\tilde{f}_\mu + \tilde{p}_r - \tilde{d}_e)_{j-1/2}$$

Here $\tilde{f}_\mu$, $\tilde{p}_r$ and $\tilde{d}_e$ denote the diffusion, production and destruction terms, respectively; they are given by

$$\tilde{f}_\mu = \frac{(1 + c_{b_2})}{\sigma} [(1 + \nu_j^+)g_j - (1 + \nu_{j-1}^+)g_{j-1}]h_j^{-1} - \frac{c_{b_2}}{\sigma} \nu_{j-1/2}^+ (g_j - g_{j-1})h_j^{-1}$$

$$\tilde{p}_r = c_{b_1}(1 - f_{t_2}) \left[\frac{(\nu^+)^2}{\kappa^2(\eta^2 + \nu_w^2)^2} f_{v2} + \sqrt{R_x}|v|\nu^+ \right]$$

$$\tilde{d}_e = \frac{(\nu^+)^2}{\eta^2 + (\nu_w^*)^2} \left[c_{w_1} f_w - \frac{c_{b_1}}{\kappa^2} f_{t_2} \right]$$

$$(\nu_w^*)^2 = \begin{cases} 0 & \text{for wall boundary layers} \\ R_x \left[1 - \left(\frac{x_{te}}{x} \right)^2 \right] & \text{along the wake} \end{cases}$$

The subscript on x, x_{te}, denotes trailing edge and the term $\frac{\tilde{p}_r}{\nu^+}$ denotes the variation of the production term with respect to ν^+; its expression depends on how the variation is accounted for. In order for the solutions to converge quadratically, the variation should be performed with respect to ν^+ on all the variables in the production term which are functions of ν^+. $\left(\frac{\partial \tilde{p}_r}{\partial v} \right)$ et al. are given by

$$\frac{\partial \tilde{p}_r}{\partial v} = c_{b_1}(1 - f_{t_2})\sqrt{R_x} \begin{cases} \nu^+ & \text{if } v > 0 \\ -\nu^+ & \text{if } v < 0 \end{cases}$$

$$\frac{\partial \tilde{d}_e}{\partial v} = \frac{(\nu^+)^2}{\eta^2 + (\nu_w^*)^2} \left[c_{w_1} \frac{\partial f_w}{\partial v} \right]$$

$$\left(\frac{\partial \tilde{f}_\mu}{\partial \nu^+} \right)_j = \frac{1 + c_{b_2}}{\sigma} \frac{g_j}{h_j} - \frac{1}{2} \frac{c_{b_2}}{\sigma} \frac{g_j - g_{j-1}}{h_j}$$

$$\left(\frac{\partial \tilde{f}_\mu}{\partial \nu^+} \right)_{j-1} = -\frac{1 + c_{b_2}}{\sigma} \frac{g_{j-1}}{h_j} - \frac{1}{2} \frac{c_{b_2}}{\sigma} \frac{g_j - g_{j-1}}{h_j}$$

$$\frac{\partial \tilde{f}_\mu}{\partial g} = \frac{1 + c_{b_2}}{\sigma} \frac{1 + \nu^+}{h_j} - \frac{c_{b_2}}{\sigma} \frac{\nu_{j-1/2}^+}{h_j}$$

$$\frac{\partial \tilde{p}_r}{\partial \nu^+} = -c_{b_1} \left(\frac{\partial f_{t_2}}{\partial \nu^+} \right) \left[\frac{(\nu^+)^2}{\kappa^2(\eta^2 + (\nu_w^*)^2)} f_{v2} + \nu^+ \sqrt{R_x}|v| \right]$$

$$+ c_{b_1}(1 - f_{t_2}) \left[\frac{(\nu^+)^2}{\kappa^2(\eta^2 + (\nu_w^*)^2)} \frac{\partial f_{v2}}{\partial \nu^+} + \frac{\partial f_{v2}}{\partial \nu^+} + \sqrt{R_x}|v| \right]$$

$$+ \frac{2\nu^+}{\kappa^2(\eta^2 + (\nu_w^*)^2)} f_{v2} \right]$$

$$\frac{\partial d_e}{\partial v^+} = \frac{(v^+)^2}{\eta^2 + (v_w^*)^2}\left[c_{w_1}\frac{\partial f_w}{\partial v^+} - \frac{c_{b_1}}{\kappa^2}\frac{\partial f_{t_2}}{\partial v^+}\right] + \frac{2v^+}{\eta^2 + (v_w^*)^2}\left(c_{w_1}f_w - \frac{c_{b_1}}{\kappa^2}f_{t_2}\right)$$

In the above equations, f_{v_1}, f_{v_2}, f_w, f_{t_2} and their variations with respect to v and v^+ are as follows

$$f_{t_2} = c_{t_3}\exp[-c_{t_4}(v^+)^2]$$

$$\frac{\partial f_{t_2}}{\partial v^+} = -2c_{t_4}v^+ f_{t_2}$$

$$f_{v_1} = \frac{(v^+)^3}{(v^+)^3 + c_{v_1}^3}$$

$$\frac{\partial f_{v_1}}{\partial v^+} = \frac{3c_{v_1}^3(v^+)^2}{[(v^+)^3 + c_{v_1}^3]^2}$$

$$f_{v_2} = 1 - \frac{v^+}{1 + v^+ f_{v_1}}$$

$$\frac{\partial f_{v_2}}{\partial v^+} = \frac{(v^+)^2\frac{\partial f_{v_1}}{\partial v^+} - 1}{(1 + v^+ f_{v_1})^2}$$

$$f_w = \left[\frac{(1 + c_{w_3}^6)}{(1 + c_{w_3}^6 g_1^{-6})}\right]^{1/6}$$

$$g_1 = rr + c_{w_5}(rr^6 - rr)$$

$$rr = \frac{(v^+)^2}{[\eta^2 + (v_w^*)^2]\kappa^2}\left\{v^+\sqrt{R_x}|v| + \frac{(v^+)^2}{\kappa^2[\eta^2 + (v_w^*)^2]}f_{v_2}\right\}^{-1}$$

$$\frac{\partial f_w}{\partial v} = \frac{\partial f_w}{\partial g_1}\frac{\partial g_1}{\partial(rr)}\frac{\partial(rr)}{\partial v}$$

$$\frac{\partial f_w}{\partial v^+} = \frac{\partial f_w}{\partial g_1}\frac{\partial g_1}{\partial(rr)}\frac{\partial(rr)}{\partial v^+}$$

where

$$\frac{\partial f_w}{\partial g_1} = \frac{c_{w_3}^6 g_1^{-6}f_w}{[g_1(1 + c_{w_3}^6 g_1^{-6})]}$$

$$\frac{\partial g_1}{\partial(rr)} = 1 + c_{w_2}(6rr^5 - 1)$$

$$\frac{\partial(rr)}{\partial v} = -\frac{(v^+)^2}{\kappa^2(\eta^2 + v_w^{*2})}\left[v^+\sqrt{R_x}|v| + \frac{(v^+)^2}{\kappa^2(\eta^2 + x_w^{*2})}f_{v_2}\right]^{-2}$$

$$\times (\sqrt{R_x}v^+)\begin{Bmatrix} 1 \\ -1 \end{Bmatrix}\begin{matrix} \text{if } v > 0 \\ \text{if } v < 0 \end{matrix}$$

$$\frac{\partial (rr)}{\partial \nu^+} = -\frac{(\nu^+)^2}{\kappa^2(\eta^2 + \nu_w^{*2})} \left[\nu^+ \sqrt{R_x} |v| + \frac{(\nu^+)^2}{\kappa^2(\eta^2 + \nu_w^{*2})} f_{v2} \right]^{-2}$$

$$\times \left[\frac{(\nu^+)^2}{\kappa^2(\eta^2 + \nu_w^{*2})} \frac{\partial f_{v2}}{\partial \nu^+} + \sqrt{R_x} |v| + \frac{2\nu^+}{\kappa^2(\eta^2 + \nu_w^{*2})} f_{v2} \right]$$

$$+ \frac{2\nu^+}{\kappa^2(\eta^2 + \nu_w^{*2})} \left[\nu^+ \sqrt{R_x} |v| + \frac{(\nu^+)^2}{\kappa^2(\eta^2 + \nu_w^{*2})} f_{v2} \right]^{-1}$$

The boundary conditions are

$$\eta = 0, \quad f = u = \nu^+ = 0$$
$$\eta = \eta_e, \quad u = 1, \quad \nu^+ = \nu_e^+ \tag{6.2.9}$$

In linearized form

$$\delta u_0 = \delta f_0 = \delta \nu_0^+ = 0$$
$$\delta u_J = \delta \nu_J^+ = 0 \tag{6.2.10}$$

Next, as before, Eqs. (6.2.4) and (6.2.6) are written in matrix-vector form, with five-dimensional vectors $\underline{\delta}_j$ and $\underline{r}_j$ for each value of j defined by

$$\underline{\delta}_j = \begin{pmatrix} \delta f_j \\ \delta u_j \\ \delta v_j \\ \delta v_j^+ \\ \delta g_j \end{pmatrix}, \qquad \underline{r}_j = \begin{pmatrix} (r_1)_j \\ (r_2)_j \\ (r_3)_j \\ (r_4)_j \\ (r_5)_j \end{pmatrix} \tag{6.2.11}$$

The 5×5 matrices are

$$A_0 = \begin{vmatrix} 1 & 0 & 0 & 0 & 0 \\ 0 & 1 & 0 & 0 & 0 \\ 0 & 0 & 0 & 1 & 0 \\ 0 & -1 & -\frac{h_1}{2} & 0 & 0 \\ 0 & 0 & 0 & -1 & -\frac{h_1}{2} \end{vmatrix} \tag{6.2.12a}$$

$$B_j = \begin{vmatrix} -1 & -\frac{h_1}{2} & 0 & 0 & 0 \\ (s_2)_j & (s_4)_j & (s_6)_j & (s_8)_j & 0 \\ (e_2)_j & (e_4)_j & (e_6)_j & (e_8)_j & (e_{10})_j \\ 0 & 0 & 0 & 0 & 0 \\ 0 & 0 & 0 & 0 & 0 \end{vmatrix}, \quad 1 \le j \le J \tag{6.2.12b}$$

$$A_j = \begin{vmatrix} 1 & -h_j/2 & 0 & 0 & 0 \\ (s_1)_j & (s_3)_j & (s_5)_j & (s_7)_j & 0 \\ (e_1)_j & (e_3)_j & (e_5)_j & (e_7)_j & (e_9)_j \\ 0 & -1 & -h_{j+1}/2 & 0 & 0 \\ 0 & 0 & 0 & -1 & -h_{j+1}/2 \end{vmatrix}, \quad 1 \le j \le J-1 \tag{6.2.12c}$$

$$C_j = \begin{vmatrix} 0 & 0 & 0 & 0 & 0 \\ 0 & 0 & 0 & 0 & 0 \\ 0 & 0 & 0 & 0 & 0 \\ 0 & 1 & -h_{j+1}/2 & 0 & 0 \\ 0 & 0 & 0 & 1 & -h_{j+1}/2 \end{vmatrix}, \quad 0 \le j \le J - 1 \qquad (6.2.12d)$$

$$A_J = \begin{vmatrix} 1 & -h_J/2 & 0 & 0 & 0 \\ (s_1)_J & (s_3)_J & (s_5)_J & (s_7)_J & 0 \\ (e_1)_J & (s_3)_J & (s_5)_J & (s_7)_J & (s_9)_J \\ 0 & 1 & 0 & 0 & 0 \\ 0 & 0 & 0 & 1 & 0 \end{vmatrix} \qquad (6.2.12e)$$

6.3 Extension of k-ε Model to Flows with Separation

The extension of the k-ε model with low Reynolds number effects to flows with separation is similar to the extension of the CS-model discussed in Chapter 4. We start with the wall boundary conditions given by Eq. (2.1.13) and edge boundary conditions given by Eq. (2.1.4). As in Section 4.0 we use the transformation given by Eq. (4.0.3) and again write the continuity and momentum equations in the form given in Eq. (4.0.4) and express the kinetic energy and rate of dissipation equations in a form similar to those given by Eqs. (2.1.4) and (2.1.5). The difference between Eqs. (2.1.4), (2.1.5) and the new ones are due to the use of a different transformation. The ordering of the eight first-order equations is done as described below.

For $j = 0$, with the first four equations corresponding to the wall boundary conditions, we write

$$f_0 = 0 \qquad (6.3.1a)$$

$$u_0 = 0 \qquad (6.3.1b)$$

$$k_0 = 0 \qquad (6.3.1c)$$

$$\varepsilon_0 = 0 \qquad (6.3.1d)$$

$$w' = 0 \qquad (6.3.1e)$$

$$f' = u \qquad (6.3.1f)$$

$$u' = v \qquad (6.3.1g)$$

Momentum Eq. (4.1.2c) $\qquad (6.3.1h)$

For $1 \le j \le J - 1$

$$k' = s \qquad (6.3.2a)$$

$$p' = q \qquad (6.3.2b)$$

kinetic energy, similar to Eq. (2.2.3) $\qquad (6.3.2c)$

$$\text{rate of dissipation equation, similar to Eq. (2.2.4)} \tag{6.3.2d}$$

$$w' = 0 \tag{6.3.2e}$$

$$f' = u \tag{6.3.2f}$$

$$u' = v \tag{6.3.2g}$$

$$\text{Momentum Eq. (4.1.2c)} \tag{6.3.3h}$$

For $j = J$

$$k' = s \tag{6.3.4a}$$

$$p' = q \tag{6.3.4b}$$

$$\text{kinetic energy, similar to Eq. (2.2.3)} \tag{6.3.4c}$$

$$\text{rate of dissipation equation, similar to Eq. (2.2.4)} \tag{6.3.4d}$$

$$u_J = w_J \tag{6.3.4e}$$

$$\lambda f_J + (1 - \lambda \eta_J) w_J = g_i \tag{6.3.4f}$$

$$x w_J \left(\frac{\partial k_J}{\partial x} \right) + \varepsilon_J = 0 \tag{6.3.4g}$$

$$w_J \left(x \frac{d\varepsilon_J}{dx} - \varepsilon_J \right) + c_{\varepsilon 2} \frac{\varepsilon_J^2}{k_J} = 0 \tag{6.3.4h}$$

where

$$k = \tilde{k} u_0^2, \quad \varepsilon = \tilde{\varepsilon} \frac{u_0^3}{x} \tag{6.3.5}$$

Next we write the difference approximations to the above equations with FLARE approximation applied to the momentum equation. The resulting non-linear algebraic system is linearized with Newton's method. The solution of the linear system is written in the form of Eq. (2.3.15) and is solved by an algorithm similar to that given by subroutine KESOLV, subsection 3.5.5.

7 Companion Computer Programs

7.0 Introduction

The accompanying CD-ROM contains four computer programs (see Section 7.5). The first one is for the CS and k-ε models described in Chapter 3. It is applicable only for turbulent flows. The second computer program is an inverse boundary-layer method applicable to both laminar and turbulent flows, with and without flow separation, as described in Chapter 4. It employs the CS model. The third computer program is the Hess and Smith panel method for computing inviscid flow around airfoils. It includes viscous effects, so that it can be coupled to the inverse boundary-layer method of Chapter 4 to compute flows around airfoils in an interactive manner as shown in the fourth computer program in the accompanying CD-ROM.

In this chapter we present sample calculations for each computer program. Section 7.1 contains five test cases for the k-ε model. They correspond to zero pressure gradient, favorable and adverse pressure gradient flows.

Section 7.2 contains sample calculations for the panel method of Chapter 5. Inviscid flow calculations corresponding to the NACA 0012 airfoil are performed for several angles of attack to predict the performance characteristics of the airfoil as well as the external velocity distribution on the airfoil and in the wake.

Section 7.3 presents boundary-layer calculations for both laminar and turbulent flows with transition location specified. The external velocity distribution is obtained from the panel method. Typical boundary-layer parameters, such as dimensionless displacement thickness, δ^*/c, and local skin friction coefficient, c_f, are calculated as a function of dimensionless chordwise distance x/c.

Section 7.4 discusses the application of the inviscid and inverse boundary layer methods to the prediction of lift and drag coefficients of the NACA 0012 airfoil discussed in the previous two sections. Both computer programs are coupled and special arrangements are introduced in order to calculate the airfoil characteristics at high angles of attack, including stall.

7.1 Test Cases for the CS and k-ε Models

There are five test cases for this computer program. They all use the notation employed in the Stanford Conference in 1968 [1]. For example, flow 1400 corresponds to a zero-pressure gradient flow. Flow 2100 has favorable, nearly-zero and adverse-pressure-gradient flow. All calculations are performed for Model = 1, 2, −1 and −2 (see subroutine 3.1.2). The predictions of four models with experimental data are given for c_f, δ^* and R_θ as a function of x in the accompanying CD-ROM.

A summary of the freestream and initial conditions for each flow are summarized below.

1. Flow 1400: Zero-Pressure-Gradient Flow
   NXT = 61, u_e/u_{ref} = 1.0, u_{ref} = 33 ms^{-1},
   c_f = 3.17 × 10^{-3}, R_θ = 3856, ν = 1.5 × 10^{-5} m^{2}s^{-1}, REF = 1

2. Flow 2100: Favorable, Zero and Adverse-Pressure-Gradient Flow
   NXT = 81, u_{ref} = 100 ft s^{-1},
   c_f = 3.10 × 10^{-3}, R_θ = 3770, ν = 1.6 × 10^{-4} ft^{2}s^{-1}, REF = 1

3. Flow 1300: Accelerating Flow
   NXT = 81, u_{ref} = 100 ms^{-1},
   c_f = 4.61 × 10^{-3}, R_θ = 1010, ν = 1.54 × 10^{-5} m^{2}s^{-1}, REF = 1

4. Flow 2400: Relaxing Flow
   NXT = 81, u_e/u_{ref} = tabulated values, u_{ref} = 1,
   c_f = 1.42 × 10^{-3}, R_θ = 27,391, ν = 1.55 × 10^{-4} m^{2}s^{-1}, REF = 1

5. Flow 2900: Boundary Layer Flow in a Diverging Channel
   NXT = 81, u_e/u_{ref} = tabulated values, u_{ref} = 1,
   c_f = 1.77 × 10^{-3}, R_θ = 22,449.2, ν = 1.57 × 10^{-4} ft^{2}s^{-1}, REF = 1

The input and output for each flow are given in tabular and graphical form in the accompanying CD-ROM. Figure 7.1 shows a comparison between the calculated results with experimental data for flow 1400 and Fig. 7.2 for flow 2100. The calculations for Model = 1, 2, −1 and −2 correspond to low Reynolds number flows with Huang-Lin and Chien models, zonal method and high Reynolds number flows, respectively.

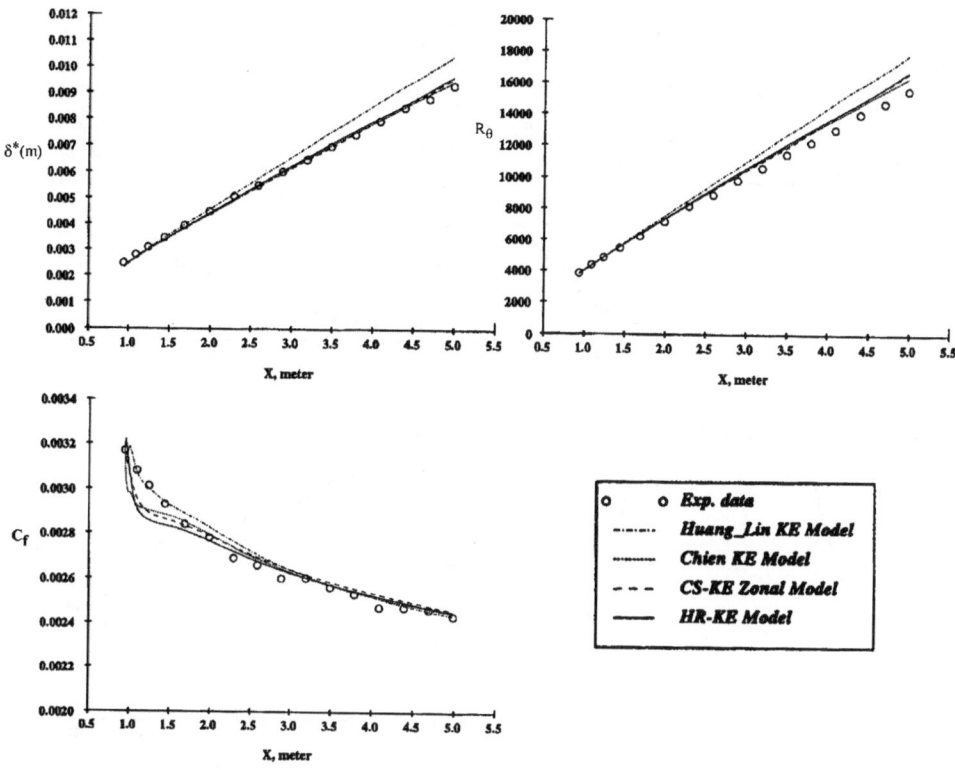

Fig. 7.1. Comparison of calculated results with the experimental data for flow 1400

7.2 Sample Calculations for the Panel Method Without Viscous Effects

This test case is for a NACA 0012 symmetrical airfoil, with a maximum thickness of 0.12c: the pressure and external velocity distributions on its upper and lower surfaces are computed and its section characteristics determined using the panel method. Table 7.1, also given in the accompanying CD-ROM, defines the airfoil coordinates for 184 points in tabular form. This corresponds to NODTOT = 183. Note that the x/c and y/c values are read in starting on the lower surface trailing edge (TE), traversing clockwise around the nose of the airfoil to the upper surface TE. The calculations are performed for angles of attack of $\alpha = 0°$, $8°$ and $16°$. In identifying the upper and lower surfaces of the airfoil, it is necessary to determine the x/c locations where $\bar{u}_e (\equiv u_e/u_\infty) = 0$. This location, called the stagnation point, is easy to determine since the $\bar{u}_e$ values are positive for the upper surface and negative for the lower surface. In general it is sufficient to take the stagnation point to be the x/c location where the change of sign $\bar{u}_e$ occurs. For higher accuracy, if desired, the stagnation point can be

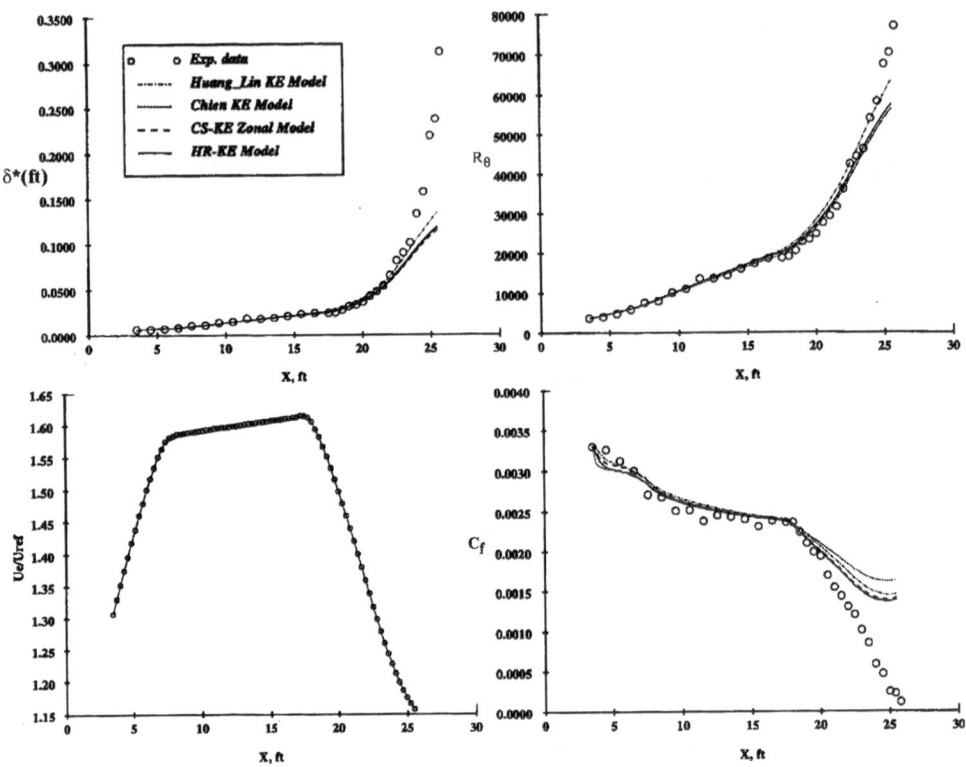

Fig. 7.2. Comparison of calculated results with the experimental data for flow 2100

determined by interpolation between the negative and positive values of $\bar{u}_e$ as a function of the surface distance along the airfoil.

Figures 7.3 and 7.4 show the variation of the pressure coefficient C_p and external velocity $\bar{u}_e$ on the lower and upper surfaces of the airfoil as a function of x/c at three angles of attack starting from $0°$. As expected, the results show that the pressure and external velocity distributions on both surfaces are identical to each other at $\alpha = 0°$. With increasing incidence angle, the pressure peak moves upstream on the upper surface and downstream on the lower surface. In the former case, with the pressure peak increasing in magnitude with increasing α, the extent of the flow deceleration increases on the upper surface and, we shall see in the following section, increases the region of flow separation on the airfoil. On the lower surface, on the other hand, the region of accelerated flow increases with incidence angle which leads to regions of more laminar flow than turbulent flow.

These results indicate that the use of inviscid flow theory becomes increasingly less accurate at higher angles of attack since, due to flow separation, the viscous effects neglected in the panel method become increasingly more im-

Table 7.1. Tabulated coordinates for the NACA 0012 airfoil

1.000000	.996060	.991140	.984290	.975520	.964880
.952400	.938140	.922150	.904490	.885240	.864460
.842250	.818680	.793860	.767880	.740840	.712850
.684010	.654460	.624290	.593630	.562610	.531330
.499930	.482486	.465056	.447665	.430339	.413103
.395971	.378964	.362108	.345420	.328917	.312618
.296550	.280736	.265190	.249928	.234965	.220333
.206040	.192102	.178538	.165366	.152604	.140264
.128362	.116914	.105932	.095430	.085421	.075921
.066938	.058480	.050557	.043180	.036365	.030116
.028319	.026575	.024883	.023245	.021660	.020130
.018656	.017237	.015874	.014568	.013316	.012120
.010980	.009895	.008867	.007894	.006977	.006116
.005310	.004561	.003868	.003232	.002653	.002132
.001667	.001260	.000910	.000617	.000380	.000201
.000078	.000012	.000012	.000078	.000201	.000380
.000617	.000910	.001260	.001667	.002132	.002653
.003232	.003868	.004561	.005310	.006116	.006977
.007894	.008867	.009895	.010980	.012120	.013316
.014568	.015874	.017237	.018656	.020130	.021660
.023245	.024883	.026575	.028319	.030116	.036366
.043183	.050557	.058480	.066938	.075922	.085424
.095432	.105933	.116916	.128364	.140266	.152607
.165370	.178541	.192106	.206043	.220334	.234966
.249926	.265191	.280738	.296555	.312622	.328918
.345423	.362109	.378968	.395977	.413111	.430347
.447669	.465060	.482490	.499930	.531330	.562610
.593630	.624290	.654460	.684010	.712850	.740840
.767880	.793860	.818680	.842250	.864460	.885240
.904490	.922150	.938140	.952400	.964880	.975520
.984290	.991130	.996060	1.000000		
.000000	−.000570	−.001290	−.002270	−.003520	−.005020
−.006760	−.008700	−.010850	−.013170	−.015650	−.018260
−.020990	−.023800	−.026670	−.029590	−.032500	−.035350
−.038180	−.040920	−.043590	−.046150	−.048590	−.050860
−.052940	−.054006	−.055004	−.055926	−.056766	−.057516
−.058179	−.058748	−.059216	−.059580	−.059836	−.059980
−.060015	−.059934	−.059734	−.059412	−.058965	−.058401
−.057710	−.056893	−.055952	−.054892	−.053715	−.052415
−.050992	−.049452	−.047799	−.046040	−.044167	−.042199
−.040134	−.037974	−.035719	−.033376	−.030954	−.028454
−.027674	−.026887	−.026093	−.025292	−.024484	−.023670
−.022849	−.022023	−.021192	−.020355	−.019512	−.018663
−.017809	−.016949	−.016084	−.015213	−.014336	−.013454
−.012567	−.011676	−.010783	−.009883	−.008977	−.008066
−.007149	−.006228	−.005303	−.004373	−.003439	−.002503
−.001565	−.000626	.000626	.001565	.002503	.003439
.004373	.005303	.006228	.007149	.008066	.008977
.009883	.010783	.011676	.012567	.013454	.014336

Table 7.1. (continued)

.015213	.016084	.016949	.017809	.018663	.019512
.020355	.021192	.022023	.022849	.023670	.024484
.025292	.026093	.026887	.027674	.028454	.030954
.033376	.035717	.037972	.040132	.042198	.044170
.046040	.047803	.049453	.050994	.052414	.053714
.054894	.055953	.056895	.057710	.058398	.058963
.059409	.059734	.059934	.060015	.059980	.059834
.059580	.059217	.058748	.058177	.057513	.056763
.055926	.055003	.054006	.052940	.050860	.048590
.046150	.043590	.040920	.038180	.035350	.032500
.029590	.026670	.023800	.020990	.018260	.015650
.013170	.010850	.008700	.006760	.005020	.003520
.002270	.001290	.000570	.000000		

portant. This is indicated in Fig. 7.5, which shows the calculated inviscid lift coefficients for this airfoil together with the experimental data reported in [2] for chord Reynolds numbers, R_c ($\equiv u_\infty c/\nu$), of 3×10^6 and 6×10^6. As can be seen, the calculated inviscid flow results agree reasonably well with the measured values at low and modest angles of attack. With increasing angle of attack, the lift coefficient reaches a maximum, called the maximum lift coefficient, $(c_\ell)_{max}$, at an angle of attack, α, called the stall angle. After this angle of attack, while the experimental lift coefficients begin to decrease with increasing angle of attack, the calculated lift coefficient, independent of Reynolds number, continuously increases with increasing α. The lift curve slope is not influenced by R_c at low to modest angles of attack, but at higher angles of attack it is influenced by R_c, thus making $(c_\ell)_{max}$ dependent upon R_c.

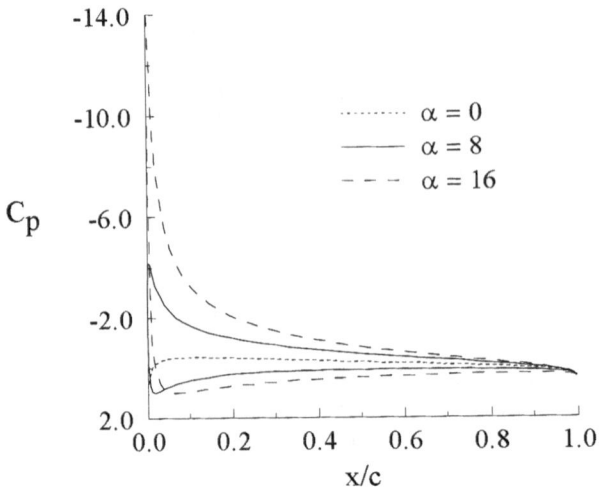

Fig. 7.3. Distribution of pressure coefficient on the NACA 0012 airfoil at $\alpha = 0°$, $8°$ and $16°$.

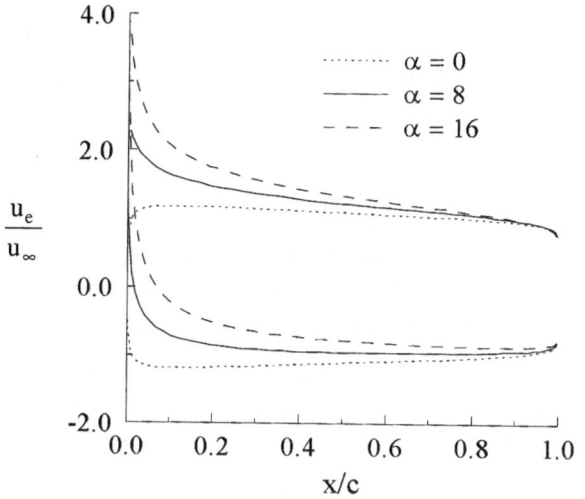

Fig. 7.4. Distribution of dimensionless external velocity on the NACA 0012 airfoil at $\alpha = 0°$, $8°$ and $16°$.

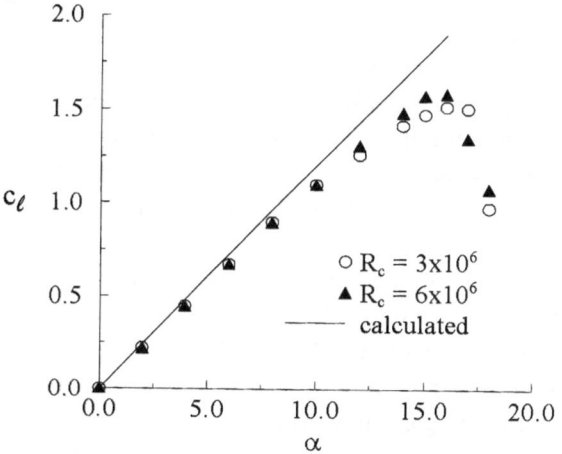

Fig. 7.5. Comparison of calculated (solid lines) and experimental (symbols) lift coefficients for the NACA 0012 airfoil.

7.3 Sample Calculations for the Inverse Boundary-Layer Program

This test case is again for the airfoil considered in the prevoius section. The boundary-layer calculations are performed only for the upper surface, for laminar and turbulent flows with transition location specified, at angles of attack of $\alpha = 4°$, $8°$, $12°$, $14°$, $16°$ and $17°$. The airfoil coordinates, x/c, y/c are used to calculate the surface distance. The calculations are done for a chord Reynolds number of 4×10^6.

In practice, it is also necessary to calculate the transition location. Two practical methods for this purpose are the Michel method and the e^n-method described, for example, in [3, 4]. The former is based on a empirical correlation between two Reynolds numbers based on momentum thickness, R_θ, and surface distance R_x. It is given by

$$R_{\theta_{tr}} = 1.174 \left(1 + \frac{22{,}400}{R_{x_{tr}}}\right) R_{x_{tr}}^{0.46} \tag{7.3.1}$$

where

$$R_\theta = \frac{u_e \theta}{\nu}, \quad R_x = \frac{u_e x}{\nu}$$

The accuracy of this method is comparable to the e^n-method at high Reynolds number flows on airfoils. The e^n-method, which is based on the linear stability theory, is, however, a general method applicable to incompressible and compressible two- and three-dimensional flows. As discussed in [3, 4], for two-dimensional flows at low Reynolds numbers, transition can occur inside separation bubble and can be predicted only by the e^n-method. For details, see [3, 4].

While the boundary-layer calculations with this program can be performed for standard and inverse problems, here they are performed for the standard problem, postponing the application of the inverse method to the following section.

The accompanying CD-ROM, under Section 7.3, presents the input and output of the calculations. Here we present a sample of them. The format of the inverse boundary-layer program is similar to the format of the interactive code and is discussed in the following section. Figure 7.6 shows the distribution of local skin friction coefficient, c_f, and dimensionless displacement thickness, δ^*/c, for several angles of attack. These results were obtained for the external velocity distribution provided by the panel method without viscous corrections. The boundary-layer calculations were performed in the inverse mode and several sweeps on the airfoil and in its wake were made. As can be seen, at low or medium angles of attack, there is no flow separation on the airfoil corresponding to the vanishing of c_f or v_w. At higher angles, however, as expected, the flow separates near the trailing edge and moves forward with increasing angle of

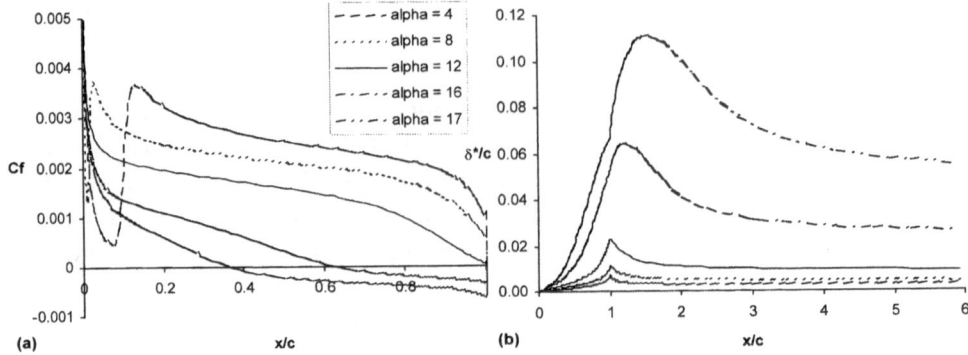

Fig. 7.6. Variation of (a) c_f and (b) δ^*/c on the NACA 0012 airfoil and its wake at several angles of attack for $R_c = 4 \times 10^6$.

attack. It is interesting to note that at $\alpha = 16°$, the flow separation occurs at $x/c = 0.6$ and at $\alpha = 17°$ at $x/c \cong 0.37$. As we shall see in the next section, interaction between inviscid and viscous results reduces the flow separation on the airfoil considerably. The results also show that, again as expected, transition location occurs very close to the stagnation point at higher angles of attack.

7.4 Sample Calculations with the Interactive Boundary-Layer Program

A combination of an inviscid method with a boundary-layer method allows the inviscid and viscous flow calculations to be performed in an interactive way. Using an inverse boundary-layer method allows similar calculations to be performed for flows including separation.

Before we present sample calculations with the interactive boundary-layer program, it is first useful to discuss the computational strategy in this program. For a specified angle of attack α and airfoil geometry $(x/c, y/c)$, the calculations are first initiated with the panel method in order to calculate the external velocity distribution and the lift coefficient. The external velocity distribution is then input to the inverse boundary-layer program in which, after identifying the airfoil stagnation point, the calculations are performed separately for the upper and lower surfaces of the airfoil and in the wake. The calculations involve several sweeps on the airfoil, one sweep corresponding to boundary-layer calculations which start at the stagnation point and end at some specified ξ-location in the wake. In sweeping through the boundary-layer, the right-hand side of Eq. (4.0.6) uses the values of δ^* from the previous sweep when $j > i$ and the values from the current sweep when $j < i$. Thus, at each ξ-station the right-hand side of Eq. (4.0.6) provides a prescribed value for the linear combination of $u_e(\xi^i)$ and $\delta^*(\xi^i)$. After convergence of the Newton iterations at each station, the summations of Eq. (4.0.6) are updated for the next ξ-station. Note that the Hilbert integral coefficients C_{ij} discussed in subsection 4.2.3 are computed and stored at the start of the boundary-layer calculations.

At the completion of the boundary-layer sweeps on the airfoil and in the wake, boundary-layer solutions are available on the airfoil and in the wake. The blowing velocity on the airfoil v_{iw} [see Eqs. (5.1.8) and (5.1.9)] and a jump in the normal velocity component Δv_i in the wake [see Eq. (5.1.10)], for which an incompressible flow are

$$v_{iw} = \frac{d}{dx}(u_{iw}\delta_A^*) \tag{7.4.2a}$$

$$\Delta v_i = \frac{d}{dx}(u_{iu}\delta_u^*) + \frac{d}{dx}(u_{il}\delta_l^*) \tag{7.4.2b}$$

are calculated and are used to obtain a new distribution of external velocity $u_{ei}(x)$ from the inviscid method. As before, the onset of transition location is determined from the laminar flow solutions and the boundary-layer calculations are performed on the upper and lower surfaces of the airfoil and in the wake by making several specified sweeps. This sequence of calculations is repeated for the whole flowfield until convergence is achieved.

The format of the input to this interactive boundary-layer (inviscid/viscous) program is similar to the input required for the inverse boundary-layer described in subsection 4.2.1. The code is arranged in such a way that it is only necessary to read in the airfoil geometry, the angles of attack to be calculated, Mach number and chord Reynolds number. The rest of the input is done internally.

We now present sample calculations for the NACA 0012 airfoil for Reynolds numbers corresponding to 3×10^6. In this case, transition locations are calculated with Michel's formula. The calculations and the results are given in the accompanying CD-ROM.

Lift, c_l, drag, c_d, pitching moment, c_m, coefficients for $R_c = 3 \times 10^6$ are shown in Table 7.2 for $\alpha = 2°$ to $16.5°$ and $M_\infty = 0.1$ together with lift coefficients calculated with the panel method. As can be seen, while at low and modest angles of attack, the inviscid lift, $c_{l_{in}}$, and viscous lift, $c_{l_{vi}}$, coefficients agree reasonably well, at higher angles of attack, as expected, they differ from each other.

Figure 7.7 shows a comparison between the calculated and experimental values of lift and drag coefficients. The agreement is good and the stall angle is reasonably well predicted. For additional comparisons with experimental data, see [3].

To describe the input and output of the computer program, we now present additional calculations for the same airfoil, this time for $R_c = 4 \times 10^6$.

The input to the IBL program (Fig. 7.8) includes airfoil geometry and/or the number of angles of attack (N), the freestream Mach number, M_∞, and

Table 7.2. Results for the NACA 0012 airfoil ar $R_c = 3 \times 10^6$, $M_\infty = 0.1$.

α	$c_{l_{in}}$	$c_{l_{vi}}$	c_d	$c_{m_{in}}$	$c_{m_{vi}}$
2.00000	0.24261	0.21099	0.00586	−0.06326	−0.04971
4.00000	0.48508	0.42567	0.00610	−0.12622	−0.10099
6.00000	0.72727	0.64337	0.00749	−0.18857	−0.15325
8.00000	0.96908	0.86241	0.00955	−0.25003	−0.20621
10.00000	1.21041	1.07109	0.01178	−0.31029	−0.25434
12.00000	1.45120	1.26253	0.01498	−0.36907	−0.29536
13.00000	1.57138	1.34396	0.01658	−0.39782	−0.31005
14.00000	1.69142	1.40836	0.01892	−0.42609	−0.31856
15.00000	1.81133	1.44754	0.02181	−0.45385	−0.31873
15.50000	1.93110	1.45653	0.02366	−0.48107	−0.31636
16.00000	1.99094	1.45811	0.02592	−0.49446	−0.31261
16.50000		1.44226	0.02837		−0.30540

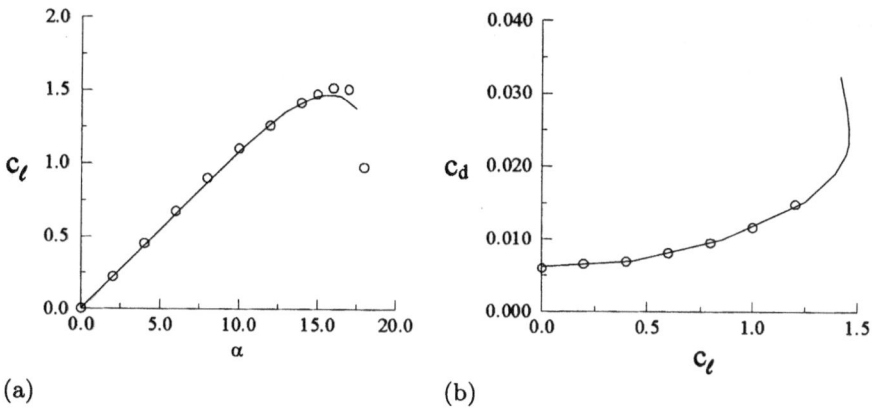

(a)   (b)

Fig. 7.7. Comparison between calculated (*solid lines*) and experimental values (*symbols*) of: **(a)** c_l vs α, and **(b)** c_d vs c_l. NACA 0012 airfoil at $R_c = 3 \times 10^6$.

the Reynolds number, R_c. The input file in the sample calculations contains the NACA 0012 airfoil coordinates which are specified by choosing either M1M4 or M1M4INP. The first choice contains only the airfoil geometry and does not contain either the angles of attack, Mach number or Reynolds number. The second choice contains airfoil geometry, angles of attack, Mach number and Reynolds number. If the first one is chosen, then it is necessary to specify N, M_∞ and R_c. For example if $N = 5$, then the angles of attack can be, say, $0°$, $4°$, $6°$, $8°$ and $9°$. Of course, these angles of attack as well as N can be changed. Then the calculations are started by specifying M_∞ and R_c. Figure 7.8 shows a sequence of the screens used for input.

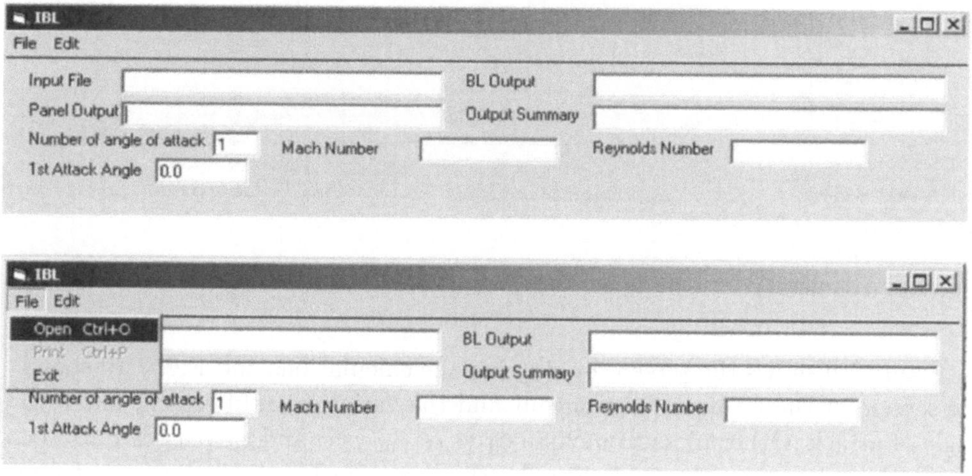

Fig. 7.8. Input format.

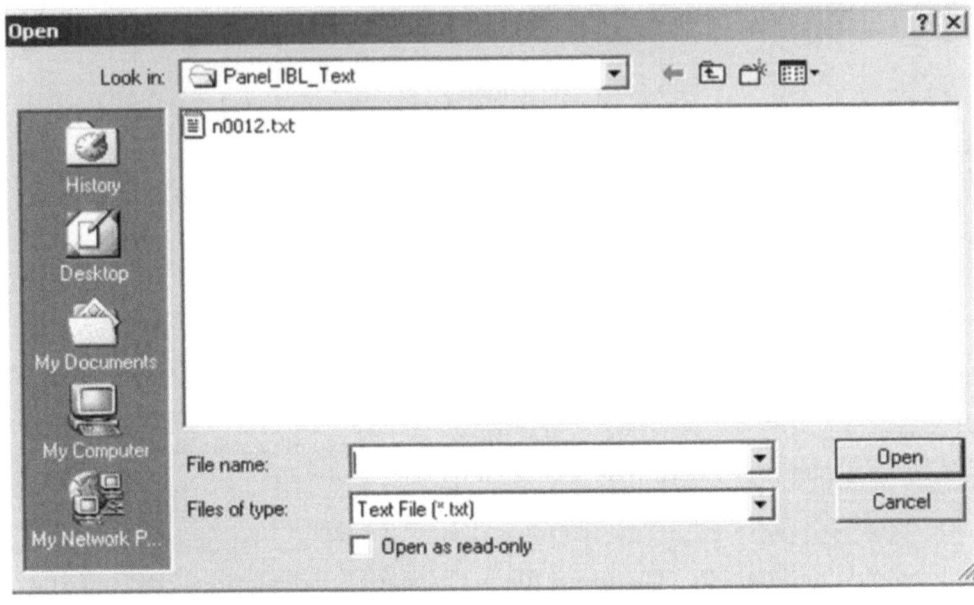

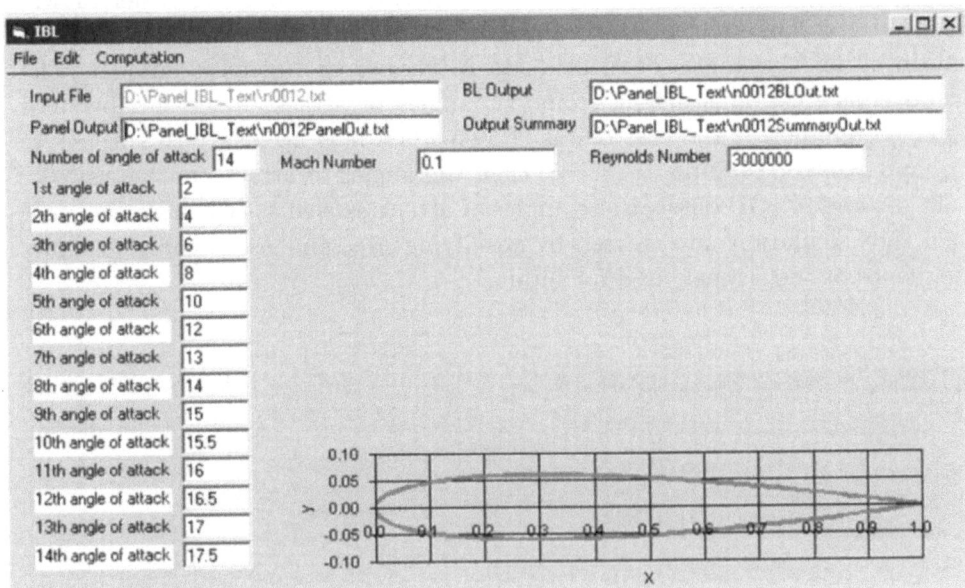

Fig. 7.8. (continued)

Figure 7.9 shows the screen for starting the calculations and Fig. 7.10 shows the screen for the format of the output and the variation of lift coefficient with angle of attack. Other plots to include c_d vs α, c_m vs α and c_d vs c_l can also be obtained as shown in Fig. 7.11. Finally, the screen in Fig. 7.12 shows that one can copy the plot to the Microsoft Word file.

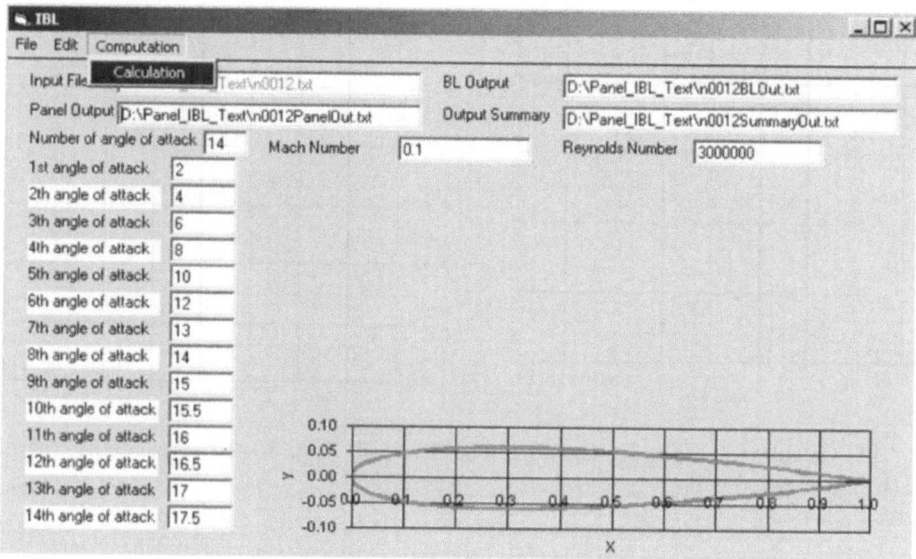

Fig. 7.9. Beginning of calculations.

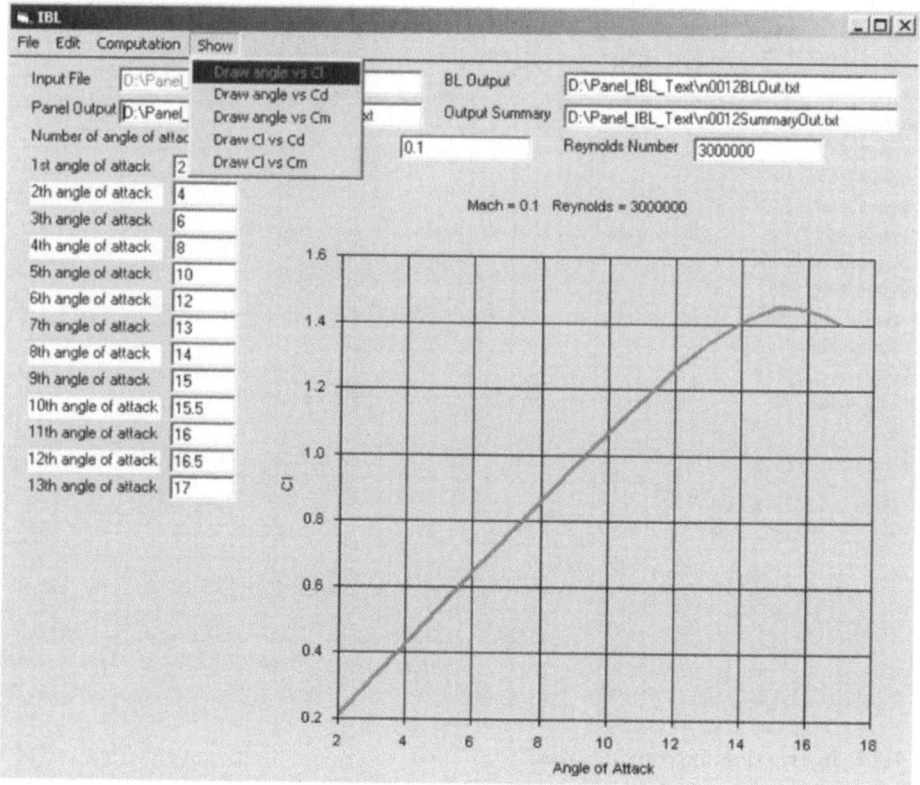

Fig. 7.10. Output format.

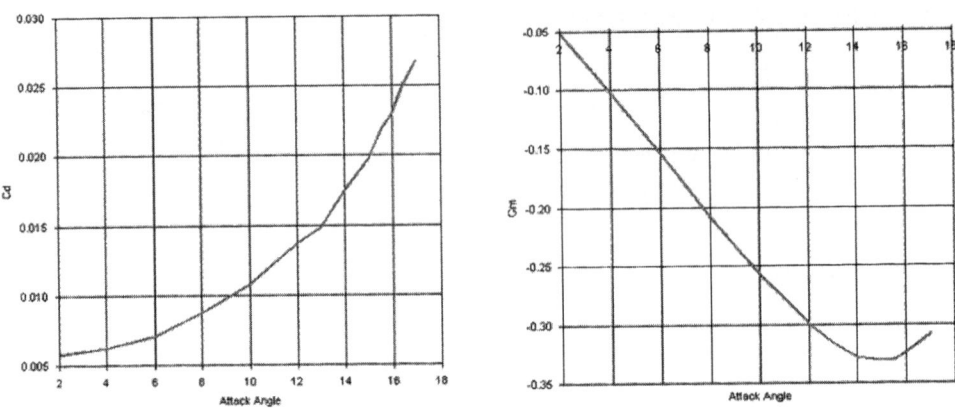

Fig. 7.11. Calculated results for the NACA 0012 airfoil, $R_c = 4 \times 10^6$, $M_\infty = 0.1$. (a) c_d vs α, (b) c_m vs α.

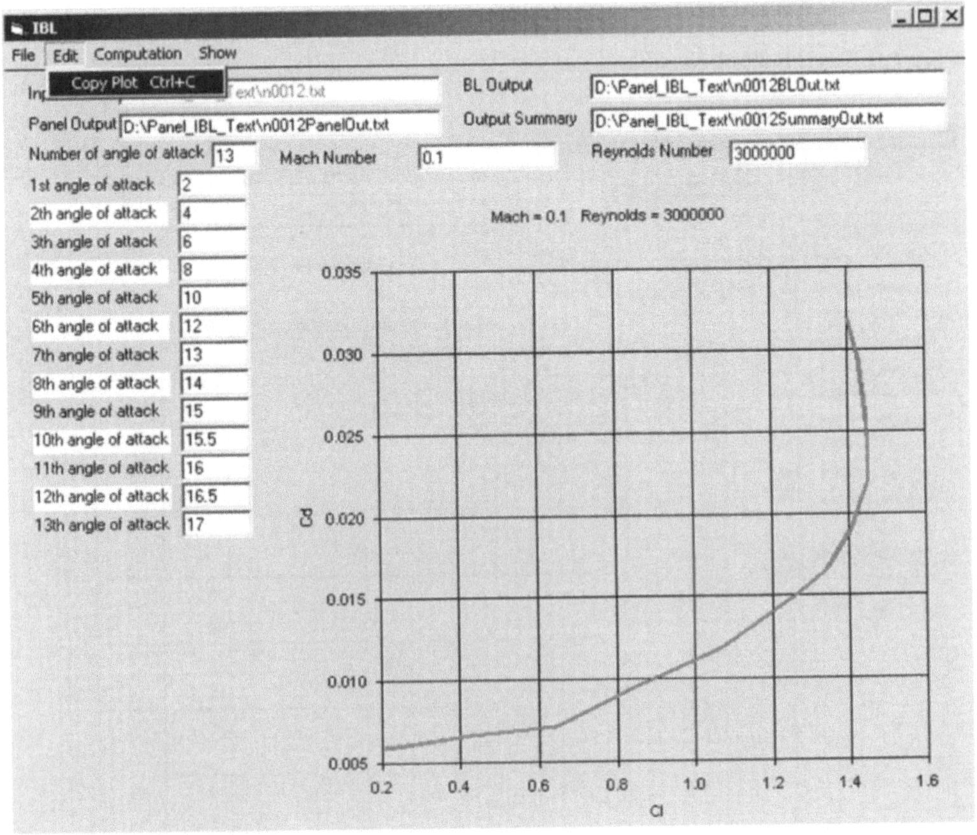

Fig. 7.12. Instruction for copying plots.

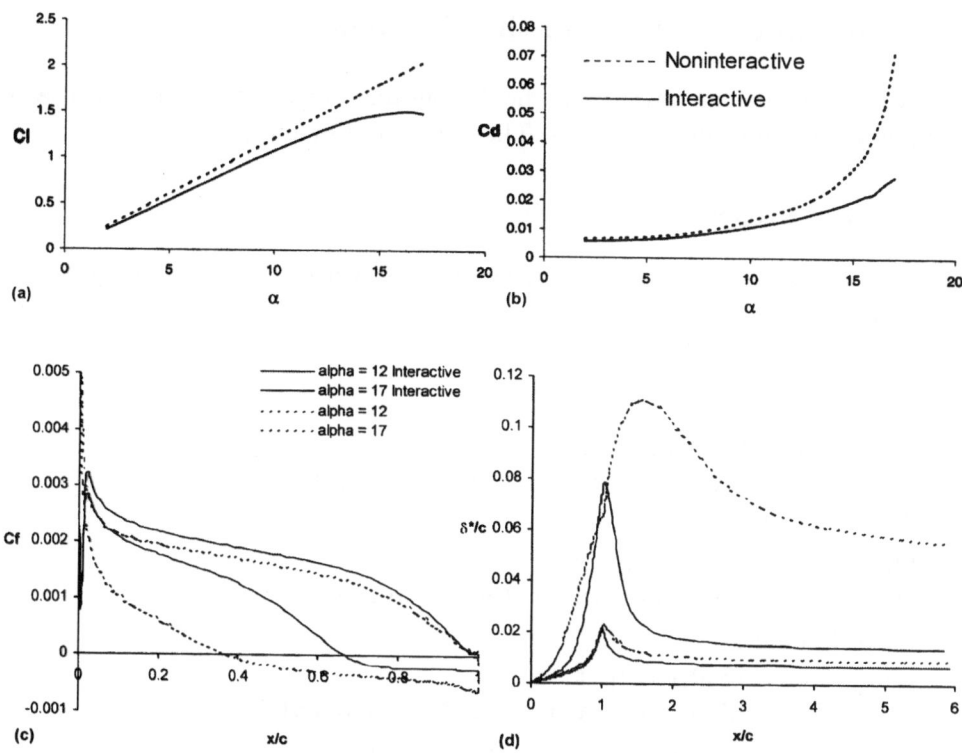

Fig. 7.13. Comparison of results between the inverse boundary-layer method and the interactive method. (a) c_l vs α, (b) c_d vs α, (c) c_f vs x/c, (d) δ^*/c vs x/c.

Figure 7.13 shows a comparison between the results of the previous section where the inviscid flow calculations did not include viscous effects, and the results of this section which include viscous effects in the panel method. Figures 7.13a and 7.13b show the strong influence of viscosity on c_l and c_d. Figure 7.13c shows that with interaction, the extent of flow separation on the airfoil decreases. For example at $\alpha = 17°$, without viscous effects in the panel method, the flow separation occurs around $x/c \cong 0.37$. With interaction, it occurs at $x/c \cong 0.62$. Similarly, with interaction, the peak in δ^*/c (Fig. 7.13d) decreases and is the reason for less flow separation on the airfoil.

7.5 Computer Programs in the CD-ROM

The CD-ROM accompanying this book contains both source and executable computer programs and test cases. They are listed below.

CS and k-ε Program

Compile `kemodel.for` (Boundary-Layer Method with k-ε Model) and generate PC executable file `kemodel.exe`.

Test Case 1300:

a) Huang-Lin Model
   Input data: `1300_1inp.txt`
   Output data: `1300_1out.txt` or `1300_1outS.pdf`

b) Chien Model
   Input data: `1300_2inp.txt`
   Output data: `1300_2out.txt` or `1300_2outS.pdf`

c) CS-KE Zonal Model
   Input data: `1300_m1inp.txt`
   Output data: `1300_m1out.txt` or `1300_m1outS.pdf`

d) Standard High-Reynolds Number KE Model
   Input data: `1300_m2inp.txt`
   Output data: `1300_m2out.txt` or `1300_m2outS.pdf`

Results Summary: `1300_out.pdf`

Test Case 1400:

a) Huang-Lin Model
   Input data: `1400_1inp.txt`
   Output data: `1400_1out.txt` or `1400_1outS.pdf`

b) Chien Model
   Input data: `1400_2inp.txt`
   Output data: `1400_2out.txt` or `1400_2outS.pdf`

c) CS-KE Zonal Model
   Input data: `1400_m1inp.txt`
   Output data: `1400_m1out.txt` or `1400_m1outS.pdf`

d) Standard High-Reynolds Number KE Model
   Input data: `1400_m2inp.txt`
   Output data: `1400_m2out.txt` or `1400_m2outS.pdf`

Results Summary: `1400_out.pdf`

Test Case 2100:

a) Huang-Lin Model
   Input data: `2100_1inp.txt`
   Output data: `2100_1out.txt` or `2100_1outS.pdf`

b) Chien Model
   Input data: `2100_2inp.txt`
   Output data: `2100_2out.txt` or `2100_2outS.pdf`

c) CS-KE Zonal Model
   Input data: `2100_m1inp.txt`
   Output data: `2100_m1out.txt` or `2100_m1outS.pdf`

d) Standard High-Reynolds Number KE Model
   Input data: `2100_m2inp.txt`
   Output data: `2100_m2out.txt` or `2100_m2outS.pdf`

Results Summary: `2100_out.pdf`

Test Case 2400:

a) Huang-Lin Model
   Input data: `2400_1inp.txt`
   Output data: `2400_1out.txt` or `2400_1outS.pdf`

b) Chien Model
   Input data: `2400_2inp.txt`
   Output data: `2400_2out.txt` or `2400_2outS.pdf`

c) CS-KE Zonal Model
   Input data: `2400_m1inp.txt`
   Output data: `2400_m1out.txt` or `2400_m1outS.pdf`

d) Standard High-Reynolds Number KE Model
   Input data: `2400_m2inp.txt`
   Output data: `2400_m2out.txt` or `2400_m2outS.pdf`

Results Summary: `2400_out.pdf`

Test Case 2900:

a) Huang-Lin Model
   Input data: `2900_1inp.txt`
   Output data: `2900_1out.txt` or `2900_1outS.pdf`

b) Chien Model
   Input data: `2900_2inp.txt`
   Output data: `2900_2out.txt` or `2900_2outS.pdf`

c) CS-KE Zonal Model
   Input data: `2900_m1inp.txt`
   Output data: `2900_m1out.txt` or `2900_m1outS.pdf`

d) Standard High-Reynolds Number KE Model
   Input data: `2900_m2inp.txt`
   Output data: `2900_m2out.txt` or `2900_m2outS.pdf`

Results Summary: `2900_out.pdf`

Panel Method (HSPM)

Compile `panel3.for` and generate PC executable file `panel3.exe`.

Test Cases NACA0012 Airfoil, $M = 0.1$

1. $\alpha = 0°$ angle of attack.
   Input file: `deg_00.inp`
   Output file: `deg_00.out` and `deg_00.bdf`

2. $\alpha = 8°$ angle of attack.
   Input file: `deg_08.inp`
   Output file: `deg_08.out` and `deg_08.bdf`

3. $\alpha = 16°$ angle of attack.
   Input file: `deg_16.inp`
   Output file: `deg_16.out` and `deg_16.bdf`

Inverse Boundary-Layer Method (IBLM)

Compile `iblm.for` (Inverse Boundary-Layer Method) and generate PC executable file `iblm.exe`.

Test Case NACA0012 Airfoil, $M = 0.1$ and $RL = 4.0M$

1. $\alpha = 0°$ angle of attack.
   Input file: `n0012a0inp.txt`
   Output file: `n0012a0out.txt` or `n0012a0out.pdf`

2. $\alpha = 4°$ angle of attack.
   Input file: `n0012a4inp.txt`
   Output file: `n0012a4out.txt` or `n0012a4out.pdf`

3. $\alpha = 8°$ angle of attack.
   Input file: `n0012a8inp.txt`
   Output file: `n0012a8out.txt` or `n0012a8out.pdf`

4. $\alpha = 12°$ angle of attack.
   Input file: `n0012a12inp.txt`
   Output file: `n0012a12out.txt` or `n0012a12out.pdf`

5. $\alpha = 14°$ angle of attack.
   Input file: `n0012a14inp.txt`
   Output file: `n0012a14out.txt` or `n0012a14out.pdf`

6. $\alpha = 16°$ angle of attack.
   Input file: n0012a16inp.txt
   Output file: n0012a16out.txt or n0012a16out.pdf
7. $\alpha = 17°$ angle of attack.
   Input file: n0012a17inp.txt
   Output file: n0012a17out.txt or n0012a17out.pdf

The results summary n0012_Out.pdf

Interactive Boundary-Layer (IBL) Program

Compile Interactive Boundary-Layer Program IBL.for (Panel and Inverse Boundary-Layer Methods) and generate PC DOS executable file IBL.exe. IBL PC Window Version is also available from this CD IBL_Window_Package.zip. Click to see installation instruction.

Test Case for NACA 0012 Airfoil, Mach = 0.10

a) $RL = 3000000.0$
   Input data: m1rm3.inp
   Output data: Panel Method m1rm3Panel.out
   IBL Method m1rm3ibl.out
   Summary m1rm3Summary.out and m1rm3Summary.bdf
b) $RL = 4000000.0$
   Input data: m1rm4.inp
   Output data: Panel Method m1rm4Panel.out
   IBL Method m1rm4ibl.out
   Summary m1rm4Summary.out and m1rm4Summary.bdf

The summary m1rm4SummaryComp.pdf

References

[1] Coles, D. and Hirst, E. A.: Computation of Turbulent Boundary Layers – 1968 AFOSR-
     IFP-Stanford Conference, vol. 2, Thermosciences Division, Stanford University, Stan-
     ford, Calif., 1969.
[2] Abbott, J. H. and von Doenhoff, A. E.: Theory of Wing Sections. Dover, 1959.
[3] Cebeci, T.: An Engineering Approach to the Calculation of Aerodynamic Flows. Hori-
     zons Publishing, Long Beach, CA and Springer-Verlag, Heidelberg, Germany, 1999.
[4] Cebeci, T. and Cousteix, J.: Modeling and Computation of Boundary-Layer Flows,
     Horizons Publishing, Long Beach, CA and Springer-Verlag, Heidelberg, Germany,
     1998.

Subject Index